FRONTLINE MANUFACTURING

Rules, Tools, and Techniques
for Line Workers

THE BUSINESS ONE IRWIN/APICS Series

Supported by the American Production
and Inventory Control Society

FRONTLINE MANUFACTURING

Rules, Tools, and Techniques for Line Workers

ROBERT A. FORCIER
MARSHA M. FORCIER

The Business One Irwin/APICS
Series in Frontline Education

Business One Irwin
Homewood, Illinois 60430

This publication is designed to provide accurate and authoritative information in regard to the subject matter covered. It is sold with the understanding that neither the author nor the publisher is engaged in rendering legal, accounting, or other professional service. If legal advice or other expert assistance is required, the services of a competent professional person should be sought.

From a Declaration of Principles jointly adopted by a Committee of the American Bar Association and a Committee of Publishers.

Sponsoring editor: Jean Marie Geracie
Project editor: Gladys True
Production manager: Bette K. Ittersagen
Compositor: Professional Resources & Communications, Inc.
Typeface: 11/14 Palatino
Printer: R.R. Donnelley & Sons Company

Library of Congress Cataloging-in-Publication Data

Forcier, Robert A.
    Frontline manufacturing : rules, tools, and techniques for line workers and supervisors / Robert A. Forcier, Marsha M. Forcier.
      p.   cm.
    ISBN 1-55623-671-9
    1. Manufacturers—Management.   I. Forcier, Marsha M. II. Title.
HD9720.5.F67   1992
670.68—dc20   92–17457

Printed in the United States of America

1 2 3 4 5 6 7 8 9 0 DOC 9 8 7 6 5 4 3 2

Dedicated to Our Faith in the American Worker

Acknowledgments

We would like to acknowledge all of the individuals at the Nelco companies and the parent corporation, Park Electrochemical Corporation, who through their pursuit of world class quality standards were an inspiration for this book. Frontline worker Eileen Theile exhibits energy that is contagious to her coworkers and her customers both inside and outside the factory. Mr. Phil Smoot, Group President and COO, provided numerous insights into team building, frontline participation, and the communication skills required to grow a results-oriented, focused organization. Ms. Emily Groehl, Corporate Vice President of Sales and Marketing, demonstrated what real customer service and attention to detail are all about. We would also like to thank Mr. Jerry Shore, Chairman and CEO of Park Electrochemical, for his pursuit of excellence, innovation, and employee involvement that have provided leadership to the entire Park family.

Our appreciation goes to Gerhard Plenert, Director of the Productivity and Quality Research Group at Brigham Young University, and to Ray Mendenhall, Quality Assurance Manager at the Clark Material Handling Company, for their valuable help during the editorial phase of this manuscript. Thanks also to Business One Irwin for their six sigma assistance—to Jean Marie Geracie, our editor, for her patience and guidance, and Editor-in-Chief Jeffrey Krames, for his willingness to take a chance on this important program.

Finally, a special thanks to our son, Joshua, a real world class guy!

Bob Forcier
Marsha Forcier

Contents

Introduction

In recent years, American manufacturing has been hit hard by foreign competition. The media report factory closings with frightening regularity. This can be viewed as a depressing situation or as a challenge for the United States to become more competitive in manufacturing. Competitiveness requires a frontline assault from workers at all levels and all positions within the factory. A first step toward becoming more competitive involves a thorough understanding of the manufacturing process. This understanding promotes confidence and leads to the production of a quality product. *Frontline Manufacturing* has been designed as a handbook to provide you, the manufacturing line workers and supervisors, with this first step.

If you are just entering the field of manufacturing, this book provides a practical background for the techniques and skills required in the manufacturing process. If you are already familiar with manufacturing, it can provide a new perspective for improving and controlling the quality of the product you are now producing.

Each chapter begins with one of 12 manufacturing rules. Each of the rules reflects the valuable experience of skilled line workers and professionals who have advanced our knowledge of manufacturing over the years. Higher yields, lower costs, higher quality products, and reduced manufacturing times are all possible with careful implementation of these guidelines.

The rules are not theories, but practical approaches that have proven successful in a variety of factories both in the United States and abroad. We encourage you to use these rules as guidelines in your everyday work environment.

In addition to the 12 rules, you will find some key words and definitions throughout the text that describe important concepts commonly used by management and skilled line workers. Knowledge of these definitions and the concepts behind them allows you to more quickly identify problems on the manufacturing line, measure and understand your process, verify that your process is in control, and systematically improve your process over time. More importantly, such knowledge will improve your communication skills. You will learn to effectively use the language of manufacturing and communicate with individuals who count—your co-workers, supervisors, managers, suppliers, and customers.

To effectively illustrate the power of the rules and definitions provided in this book, you will be introduced in Chapter 1 to a fictitious factory—Everest Mountain Bikes, Inc., which will function as a practical model. Everest is not unlike many other companies in the world today. Although small by most standards (74 employees), Everest's operation will give you a clear understanding of the inner workings of a typical manufacturing operation. Everest's employees share many of the difficulties and frustrations experienced by most factory line workers. By the end of the book, Everest will not be a perfect company—no company will ever be. However, you will be able to observe how the practical application of the 12 manufacturing principles improves Everest's manufacturing processes and consequently the company's ability to compete and survive in the marketplace. By viewing Everest's successes and failures firsthand, you will learn the latest manufacturing techniques, terminology, and methods employed by the top industrial companies of the world. You will learn how to control processes, build quality products, and improve your work environment on a daily basis.

This book is about utilizing your talents—especially your intellect—to produce quality products, and help the United States remain competitive in the world market—an important goal for us all.

Bob Forcier
Marsha Forcier

About APICS

APICS, the educational society for resource management, offers the resources professionals need to succeed in the manufacturing community. With more than 35 years of experience, 70,000 members, and 260 local chapters, APICS is recognized worldwide for setting the standards for professional education. The society offers a full range of courses, conferences, educational programs, certification processes, and materials developed under the direction of industry experts.

APICS offers everything members need to enhance their careers and increase their professional value. Benefits include:

- Two internationally recognized educational certification processes—Certified in Production and Inventory Management (CPIM) and Certified in Integrated Resource Management (CIRM), which provide immediate recognition in the field and enhance members' work-related knowledge and skills. The CPIM process focuses on depth of knowledge in the core areas of production and inventory management, while the CIRM process supplies a breadth of knowledge in 13 functional areas of the business enterprise.

- The APICS Educational Materials Catalog—a handy collection of courses, proceedings, reprints, training materials, videos, software, and books written by industry experts...many of which are available to members at substantial discounts.

- *APICS The Performance Advantage*—a monthly magazine that focuses on improving competitiveness, quality, and productivity.

- Specific industry groups (SIGs)—suborganizations that develop educational programs offer accompanying materials, and provide valuable networking opportunities.

- A multitude of educational workshops, employment referral, insurance, a retirement plan, and more.

To join APICS, or for complete information on the many benefits and services of APICS membership, **call 1-800-444-2742** or **703-237-8344**. Use extension 297.

FRONTLINE MANUFACTURING

Rules, Tools, and Techniques for Line Workers

Chapter One
Pride of Workmanship

Rule One

Maintain pride of workmanship, and manufacture only the highest quality products.

T he underlying theme in all successful manufacturing—and the subject that will be repeated throughout this book—involves quality. Without quality no one will want your products; and without investing your personal quality, you won't enjoy your jobs. Quality and pride of workmanship are the cornerstones of successful manufacturing. Both of these elements can be achieved only through a commitment from you, the line worker or supervisor.

Everyone has experienced pride at one time or another. Pride is the self-satisfaction that comes from accomplishing something—cooking a great dinner, fixing a car, passing an exam at school, making a discovery at work. When we perform work successfully around the house, at school, or on the job, it gives us a sense that we have achieved the goals that we have set for ourselves.

In the workplace we don't always get recognition for a job well-done through a pat on the back or some other perk. It would be nice to receive recognition from outside sources every time we did a good job, but life just doesn't always work that way. We must recognize that in our competitive world putting forth our best effort everyday helps ensure the quality of our products, which in turn helps ensure our job security. We can gain and maintain world competitiveness through quality products. When we know we have put forth our best effort, we can then give ourselves a pat on the back. Generally, other forms of recognition will follow.

Quality and pride of workmanship can actually be *seen* in many different products. Good examples of this are the many fine paintings and other artifacts that were produced hundreds of years ago. They maintain an ageless quality because the craftspeople who produced them took great pleasure in using their natural skills and talents. These artists probably earned a livelihood from the creation of their products, but it is obvious that quality and pride go far beyond monetary gain. When self-satisfaction is invested in a job, it won't cost anything, but the payoff in quality will ensure future employment.

Recently, American workers have been attacked by foreign competitors who claim that Americans don't work hard enough. These competitors are completely wrong—Americans have always been hard workers and have always associated hard work with personal achievement. We didn't get where we are by sitting on our hands!

Where we might have missed the mark recently is that American workers don't feel that their efforts count. Sometimes we just do what is necessary to get by rather than using our natural talents and skills to produce a quality product. It is becoming increasingly evident that our efforts do count. In order to survive, American manufacturing needs more than your physical strength and stamina. It needs your skills and talents to produce the best products on the market. You are the key ingredient, and the fact that you are reading this book indicates that you are willing and more than able to accept the challenge of world competition.

Things are changing, and you are a big part of this change. Today, hard work alone does not always lead to success. In this book we will explore the latest techniques in working *smarter*. You've probably noticed that studying laboriously for an upcoming exam doesn't always yield a good grade. However, you can usually achieve high scores if you study *intelligently* for the tests. It is the method by which we approach the job in front of us that determines our effectiveness. In this book we will explore many techniques that will improve your effectiveness. You will gain an edge through the use of your intelligence. You will have the opportunity to shine when customers clamor for your quality products. This is called pride of workmanship.

> **Pride of Workmanship** The new American work ethic that emphasizes high quality products, intelligent manufacturing, and a spirit of involvement by all workers in the manufacturing process and product.

Intelligent manufacturing requires that we work smarter and thereby balance our physical strength and dexterity with sound decision making, innovation, and problem solving.

Simply stated, the intelligent use of various manufacturing tools can significantly reduce labor content and improve product quality. These manufacturing tools may include hand tools, machinery, and vehicles. Still others are planning and control tools that chart and measure the results of the manufacturing process. By utilizing a wide variety of tools in an effective manner, you can complement your physical strength and build quality products. Working smarter makes the manufacturing process more interesting and less tiring than the older concept of simply working harder to produce quality goods.

Pride of workmanship is also about involvement—in your product, with your fellow employees, your suppliers, and your customers. The greatest pride comes when all the workers in the manufacturing team understand the importance of quality product and working together toward the same goal. Get involved in your work, your product, and your business. Don't waste your time on a meaningless job. Give your job some value. Perform your work so that all can see your pride.

QUALITY

There can never be enough said about quality and its pursuit. Each step in the manufacturing process, starting with the inception of the original design and continuing through the delivery of the final product to the customer, is an opportunity to inject quality.

Quality is a yardstick used to measure your work—a method you can use to test your progress and the product you are building. This concept of quality is simple and should remain clear in your mind when you perform your various functions in a manufacturing facility. Let's look at the key words in the definition.

Quality   A measure or degree of excellence.

Although this is the shortest definition you will come across in this book, it is packed with important information. You've evaluated items in your personal life in terms of quality. You can see quality in certain products—a new appliance, a garden tool, or an automobile. If you observe carefully, you can see (or measure) that a particular product works better or lasts longer than another similar product. Based on this observation you probably make your decision on which products to buy.

In manufacturing, the ability to see quality in products is enhanced through the use of some measurement tools. Most often quality is measured in terms of statistics. Statistics will be discussed in detail later, but basically we measure something statistically by comparing it to other similar products.

As you learn more about the skills of manufacturing, you will understand and rely upon a variety of measurements including statistics that will provide invaluable feedback on your performance and your ability to satisfy your customers. Part of achieving excellence in manufacturing involves precisely measuring your progress along the way. By visually scanning, listening to, and feeling the product you are manufacturing, you can certainly measure your progress. However, with statistics and other accurate measurement tools, you will learn how to improve your manufacturing process and see true quality with a clarity not possible through observation alone.

Quality Leaders

Several individuals have stressed the importance of quality in the manufacturing process, and most agree that improved quality does not necessarily mean increased manufacturing cost. In fact, increased quality in the manufacturing process can actually lower waste and lower cost. Following is a list of those most often credited with conveying the spirit of quality in manufacturing:

Philip B. Crosby   A leading American quality expert and consultant whose book *Quality is Free* was a landmark writing on the importance of quality in American manufacturing, Crosby emphasized that quality products improved profits and reduced costs.

William Edwards Deming   Credited as a major contributor to Japan's manufacturing prowess, Deming encouraged the use of statistical methods to measure quality manufacturing. Deming began his career working for the Census Bureau during WW II with statistical quality control methods and then worked with the Japanese after the war. The Japanese later established the *Deming Prize* awarded to companies that excelled in the use of statistical methods.

J. M. Juran   One of the first quality experts in United States manufacturing, Juran is the author of the classic *Quality Control Handbook*. Juran is also noted for his early work in Japan and for starting the Juran Institute— a school for quality control.

The ideas of Deming and Juran originated in the United States but gained a stronger foothold in Japan. Their concepts included the use of statistical methods, which became so powerful that they formed the foundation for Japan's impressive rise in manufacturing expertise. Deming and Juran proved through numerous success stories in many factories throughout the world that a strong emphasis on quality contributes to company growth, greater profits, and strong customer satisfaction. We have learned from Crosby, Deming, Juran, and others the importance of quality in manufacturing. We have also learned to *measure* progress and reward certain companies for their success in the quality area. Deming's influence in Japanese business was so powerful that the Deming Prize, which is as prestigious in Japan as the Oscar is in the United States, was established to recognize exceptional companies. Japan has emphasized quality through various awards, but the United States is not standing still in this area. The Malcolm Baldrige Award, established in 1987, which recognizes the substantial efforts of American companies, is fast becoming *the* prize in the United States, and is receiving global attention. American recipients of this prestigious honor richly deserve the recognition and attention they receive.

Malcolm Baldrige National Quality Award

This quality award is given to American companies for outstanding quality performance. Initiated by an Act of the United States Congress, the award is dedicated to the late Malcolm Baldrige, former Secretary of Commerce.

If your company is involved in competing for the Malcolm Baldrige Award, it is important to remember that the winning or losing of this award is not the most important order of business. The skills, attitude, and quality concepts described in the Malcolm Baldrige Award are what really count. At the heart of the award are quality, pride of workmanship, and a strong focus on customer satisfaction. The Baldrige Award and others are important to our profession because they recognize the process of improving manufacturing skills.

Quality and pride of workmanship are not restricted to the United States and Japan. There have been some outstanding successes in many other countries where quality concepts are pursued. A special term has been coined to describe the global acceptance of quality and pride of workmanship.

World Class Manufacturing (WCM)   High quality manufacturing skills and products that are competitive on a global basis.

Imagine taking a whole company, including its factories and personnel, and placing it anywhere in the world. Imagine further that this same company *immediately* started producing product for the local market, and that the local market could actually see the same quality.

World Class means exactly that—that you have a *classy operation* that can compete anywhere in the world because the people, product, and technology are outstanding, and the quality of the products can be seen by anyone. How does this happen? It happens by starting with pride of workmanship and manufacturing quality products.

Reliability

We have emphasized the importance of the quality that can be seen in your product. Will that quality last over the product's lifetime, or will the product break down or fall apart? If we build a high quality product, it should stand the test of time. If it doesn't, time, money, and resources have been wasted. Quality, measured over time, has a special name.

Reliability   The measure of how long a product will last in the field without failing or requiring repair, reliability is a general measure of quality over a period of time.

When consumers invest significant money in a product, they want it to LAST. There is a direct relationship between how well a product is built and its ultimate reliability. Generally, the higher the quality level, the longer the product will last in the field. Your goal should be to produce products that are durable and withstand the test of time.

EVEREST MOUNTAIN BIKES, INC.

In order to provide a working example of quality, pride of workmanship, and other key concepts in manufacturing, we have created an imaginary company of 74 employees—Everest Mountain Bikes, Inc. Everest will be our vehicle for exploring some of the relevant problems and opportunities you will encounter in your career in manufacturing. It will also offer a glimpse of the world of management, and how it operates. Everest Mountain Bikes, Inc. is a modern company with a successful product and a strong desire to stay in business.

Since the mountain bike market is highly competitive, Everest is always seeking ways to gain an advantage. The employees and the managers want to produce products that spark their customers' interest and excitement. Their products range from the highly successful X6 deluxe mountain bike to the newly developed 400T tandem mountain bike. Throughout this book, you will see Everest succeeding and failing in its manufacturing approach and be able to learn from the company's experience.

Everest has a mix of employees with conflicting ideas on how to make a bicycle lighter, with higher quality and more profit. Why are there so many differing opinions? Bicycles should be easy to manufacture; they certainly don't have the complexity of manufacturing that an automobile or a computer system does. The same manufacturing principles apply to the manufacture of bikes, cars, computers, or anything else.

Everest is comprised of a variety of employees including Eileen, a frontline worker in Wheel Assembly who is progressive, proactive, and ready to conquer any new challenge that comes her way, and Rick, an excellent contributor in the Frame Assembly department. Additionally, Jaime, Mike, and others throughout the plant will illustrate interesting problems and opportunities that arise in the Everest factory.

The tightrope act in manufacturing is to make customers, employees, and suppliers satisfied winners. Everest faces many of the same problems encountered by a number of American manufacturers. For example, Everest has foreign competition that is shipping good product into the United States with 30 percent lower pricing. Sound familiar? Everest also faces the daily frustrations of equipment breakdowns, tool problems, and communication problems. These can be serious since Eileen's, Rick's, and other employees' jobs are on the line. The strength and security of the Everest employees involve their ability to work as individuals and as a team. They must utilize a variety of rules, tools, and techniques to improve their competitive edge. The diversity of the Everest work force may be the company's greatest resource.

TIPS

(1) Assure that every time you do your job, regardless of the task you are performing, you are doing the best job you can possibly do. The quality you put into your manufacturing process will ensure your job. For example, if you are on an assembly line fastening a body part to an automobile, assure that all the

fasteners or welds are correctly installed and located. The completed assembly should be structurally sound, rattle-free, and ready for the next operation. If you are in an electronics assembly operation, assure that all the components are placed correctly, the soldering is defect-free, and the assemblies are cleaned after your process is complete.

(2) It is important to occasionally refresh your quality perception. This can be accomplished by visiting a showroom of automobiles or appliances where you can observe more than one brand of product. Carefully observe the details in comparable products. Determine if you can *see* the quality in the products and see if there are quality differences among the brands displayed.

(3) If you observe an operation or a product in your factory that needs quality improvement, speak up. Get involved in your manufacturing process.

CHAPTER REVIEW

(1) Name two items or accomplishments that you have taken pride in during the past year. These can be things that you have made or services you have provided. What does pride feel like?

(2) Quality is a measure of what?

(3) Which two American quality experts helped Japan become an economic power by using statistical methods to improve quality?

(4) What is the most important quality award in the United States?

(5) What term is used to describe a company that understands quality and is competitive on a global basis?

(6) Name two well-known United States companies that compete successfully on a global basis.

(7) Look up ISO 9000 in the Glossary. How would this standard relate to world class manufacturing?

Chapter Two
Make It Right the First Time!

Rule Two

Practice the discipline necessary to manufacture the product correctly the first time.

Knowledge of the manufacturing process helps you perform your job correctly the first time, every time. It is important to learn *why* your job requirements exist, and what purpose they serve. In this chapter we will discuss where to get this information, and how to use it to "make it right the first time."

Whether you are dealing with mountain bikes, automobiles, or donuts, there is a body of knowledge that describes the manufacturing process and procedure. That knowledge explains how and why manufacturing processes take place.

As a line worker or supervisor, it is your responsibility to gain as much of this knowledge as possible through a variety of sources. Usually this knowledge is provided by management through various training tools. By tracking down this information, you will understand more clearly where your job fits into the big picture, and why management makes decisions the way it does.

With this knowledge, the entire manufacturing process will make more sense. However, rather than concentrating on the management view, we will take only a few snapshots of management, as necessary. A company's Mission Statement, described below, is such a snapshot. The Mission Statement is a management tool that provides basic direction for the company and furnishes valuable insight into a company's inner workings.

THE MISSION STATEMENT

All businesses want to make money. The generation of profits sustains a business and allows it to grow and prosper over time. How a business generates profits defines the nature of its activities. More and more companies are developing policy statements that, if followed, help ensure that they will remain competitive and make money. These statements usually give direction to the business and are referred to as Mission Statements.

> **Mission Statement**   A concise one-page description of a company's purpose and quality philosophy, a Mission Statement is distributed widely to customers, suppliers, and employees.

Everest, our imaginary company, decided during its second year in business to develop a Mission Statement. As with anything worthwhile, a Mission Statement takes time and effort to develop. Everest management decided that its employees and managers should meet on a regular basis to develop a Mission Statement, thereby allowing everyone to participate in the company's future direction.

Employees were given a starting point for their Mission Statement: Everest bikes are special since they are used for the specific application of climbing hills and cross-country trail riding. They are considered *high-end* bicycles—high quality, high performance bikes. After several sessions, Everest employees hammered out the following statement:

Everest Mountain Bike, Inc. Mission Statement

Our Company   Everest is, and will continue to be, a world class manufacturer of high quality mountain bicycles and is dedicated to its customers, its employees, its products, and its technology.

Our Commitment   Each Everest employee will strive to manufacture the Everest products correctly the first time and maintain high quality standards by following procedures and specifications throughout the factory. Pride of workmanship and quality will be the cornerstones of the Everest manufacturing team.

The Bottom Line   Everest will remain profitable in the long term through the development of new mountain bike products, minimizing factory waste, and continuous improvement of the entire manufacturing process.

This statement gives Everest a purpose or a premise, and makes it clear that mountain bikes and Everest are synonymous. It also provides a clear direction for both line workers and supervisors by emphasizing certain key phrases such as quality, pride of workmanship, and manufacturing the product correctly the first time.

Imagine the power that lies in the hands of those who know where they are going! The Everest Mission Statement provides the first step in the *know-how* of the company. Your company may or may not have a Mission Statement, but most companies have a direction that is communicated to their employees in some type of written format. It is important to know this direction since it provides an explanation for the existence of your job. Take the time to learn your company's Mission Statement or its direction. It will give you a broader view of your job function.

THE MANUFACTURING PROCESS

The next step in understanding your company is to learn the manufacturing process. In Everest's Mission Statement, we read that Everest is a world class manufacturer of mountain bikes. There is a distinct value in manufacturing mountain bikes that eventually return profits to Everest. Adding value is the primary purpose of any manufacturing operation. In fact, manufacturing generates wealth to society by providing jobs and generating a flow of currency among companies and individuals.

This value comes from carefully designing and delivering a product that can be useful. If our customer views our product as valuable in comparison to other available products, we will be paid fairly and equitably. If we do not add value—as perceived by the customer—the cycle will eventually end because the customer will no longer pay the price. It sounds simple, and it is simple. To achieve value in each product we build, we need a plan of what we want to build (*design*); we need to build it from something (*material*); and we need to combine the two with a procedure for building it (*process*).

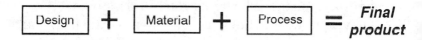

In any factory the formula remains the same. A design, a material, and a process combine to make a product for a customer. The manufacturing know-how that makes a company successful resides in these three separate pockets. For example, workers in the Paint department at Everest learn the steps to follow in the painting operation. They learn what a painted bike is supposed to look like, and what procedures they are to follow when painting. This is where labor can be turned into pride of workmanship and quality products.

Learning Your Process

Typically, you will learn your job and its associated design, material, and process requirements through either a coworker, a supervisor, or specialized training. You will pick up the knowledge you need through this training. Whenever you are offered the opportunity for additional training, take it—it is always for your benefit. Increased skills and knowledge make you a more valuable employee.

Company knowledge of processes and procedures is not always completely written down in an easy to understand fashion. Not all companies have formal, organized training programs. If yours doesn't, more responsibility is placed on you to seek out the information you need to perform your job. The design, material, and process information will probably exist in a variety of formats: drawings, training materials, documents, and manuals.

Training may be provided by a fellow employee who trains you at your work station (referred to as *on-the-job training*), or a manager who trains you in a classroom setting. If you have questions about your operation, ask them. If you need more training in your area, ask for it. Read the documents and drawings given to you. Knowledge and skill are your greatest assets in the manufacturing arena.

In order for you to manufacture your product correctly the first time, you must know what is expected of you. If you don't, your quality output will be poor, and you may have to repeat your work—which may jeopardize your job and eventually the company. Learning the three components of manufacturing (design, material, and process) through a training process improves your manufacturing skills and provides you with strong problem solving skills.

Design, material, and process are connected and influence each other. Some designs are not compatible with certain materials, and some materials are not compatible with certain processes. The right combination of these elements yields a valuable product. Even a slight deviation from the right combination may yield a defective product. As you become skilled in manufacturing, you will gain a feel for the right combinations. Analyzing what works and what doesn't helps to improve your problem solving skills.

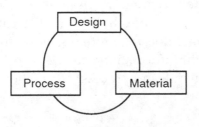

At Everest, for example, an experienced painter would understand that an epoxy paint acts differently than an acrylic paint during the spraying operation. Acrylic paint requires a slightly different air pressure and spray gun setting to achieve a smooth, defect-free finish. When acrylic is a requirement of the design, the painter knows to adjust his or her spray gun to compensate for the different material and achieve quality results. Knowing when to adjust the sprayer is what makes the painter experienced and valuable to the company. The painter possesses knowledge that helps to ensure his or her job.

Many manufacturing people do not clearly understand the interrelationship of design, material, and process. If you are beginning to understand, you are well on your way to a successful career in manufacturing. This method of analyzing the design, material, and process is a powerful tool that you can use daily.

By observing how design, material, and process affect each other, you will be able to solve many of the problems that occur in the factory. Humans make mistakes. Mistakes can be an opportunity to learn. However, mistakes in a factory can be costly and should be kept to a minimum. Your company is likely to tolerate a mistake the first time but will probably take a dim view if you make the same mistake twice. Keep a personal work journal. When you make a mistake, analyze why and how it occurred. Write it down for future reference. Was the material incompatible with the design or process? Make notes. Don't make the same mistake twice.

Following Procedures

Manufacturing growth is a positive sign and usually means success, profitability, and continued operations. However, companies can easily get into trouble. Everest, in fact, ran into trouble last fall at a time when its business was good, and the company was growing at about 10 percent a year. Everest developed a capacity problem. Rick, a long-term employee in the paint section of the Frame Assembly department, got caught in the middle.

Orders were pouring in for a new tandem bicycle, the 400T, and work was piling up. Rick, who was responsible for painting and curing the bicycle frames, mentioned to his department manager that there was a capacity problem in his department. The department only had two ovens for curing the bicycle frames. As sometimes happens, management did not adequately address the problem. Rick felt obligated to keep up with production. The pressure increased steadily.

Suddenly, Rick had a brainstorm. He remembered that the company used acrylic paints on a batch of bikes in the past. He also remembered that the oven was 20 degrees higher (260 to 280 degrees Fahr. instead of the present 240 to 260 degrees Fahr.), and the curing process was one hour shorter. The company was using epoxy paint now instead of the acrylic because it yielded a glossier, more durable finish. Rick thought that it might be a good idea to turn up the oven temperature for the bikes with the epoxy finish to shorten the cycle and relieve the production bottleneck. He discussed this change with his supervisor and some coworkers. They thought that it was worth trying.

Rick ran a test the next day, and it worked great. The paint cured beautifully. Rick cured all the bikes with the new process. Within three days, he caught up with his work. Six days later, however, disaster struck. Someone in shipping noticed that a three-inch strip of paint had peeled off the fork of a bike that was ready to ship to a customer. Rick researched the problem and discovered that it was one of the first bikes baked at the higher temperature.

Rick had changed an Everest manufacturing process without understanding the implications of his actions. He failed to realize that his change to the process did not fit the material or design of his product. What worked with the acrylic paint did not work with the epoxy paint. While Rick had performed a test run—which was good—it was not of sufficient length to detect the problem.

The epoxy paint came with specifications from the manufacturer stating that baking above 260 degrees Fahr. required a special primer for proper adhesion of the paint. The specifications further warned that this was especially necessary when coating aluminum. Rick was not aware that Everest had changed not only the paint type on the new tandem model, it had also changed the frame material from steel to aluminum to save weight. Rick's new process of baking at the higher temperature resulted in an incompatible combination of aluminum and epoxy paint. Everest had defective products.

Ultimately, the defective products were reworked, costing the company additional time, money, and material. Rick and his coworkers failed to recognize the interrelationship of design, material, and process. Also, they did not follow the manufacturer's specifications and company procedures.

INSIDE THE GOAL POSTS

Unfortunately, Rick learned a lesson the hard way about making it right the first time. He did not follow the Everest procedure. Whether it is mountain bikes or automobile manufacturing, following procedures and specifications is fundamental. What exactly are the specifications and procedures, and where do we find them?

Generally, companies maintain formal documents called *Specifications* and *Procedures*—the recipes to be followed for successful manufacturing. The following are commonly used terms that mean the same as specifications or procedures.

Specification Terms	Procedure Terms
Standards	Shop Practices
Tolerances	Travelers
Customer Requirements	Standard Routings
Customer Specifications	Manufacturing Instructions
Engineering Drawings	Methods
Engineering Specifications	

Specifications and procedures may be found in computers, on purchase orders, on engineering drawings, or in instruction manuals.

Specifications Detailed descriptions of dimensions, materials, quantities, temperatures, and so on, used to manufacture product.

Notice that a specification requires measurement. A certain amount of steel is measured to make an automobile fender. A painter measures very specific amounts of pigment to produce a certain color. Rick, at Everest, cured the paint at a certain temperature. The temperature was a measurement of heat. Each manufacturing process requires specific measurements, and the specific measurements for each process are found in the specifications. Specifications are developed through factory testing or customer requirements.

Specifications exist for each design, material, and process. As stated earlier, it was acceptable for Rick to cure the epoxy paint in the 240 to 260 degree Fahr. range. In most manufacturing processes there is an acceptable operating range. Measurements within this range will yield good product, measurements outside of this range yield defective product.

The technical term for these practical limits, which will be used many times in your manufacturing career, are the *upper* and *lower specification limits*. These limits create a manufacturing window or operating range. A good way to visualize the meaning of a specification limit is to think of two goal posts that have been set up for each operation.

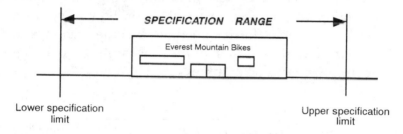

Lower specification
limit

Upper specification
limit

With careful planning and preparation, you can go through the goal posts (specification limits) each time you perform an operation. Positioning yourself (and your factory) within the goal posts indicates that you are within the requirements set down by either the company or your customer. Failure to keep your manufacturing process between the goal posts will result in defective product. You will be *out of spec*.

If you can keep your activities centered within this window, you will produce a defect-free product. You will hit your goals and make it right the first time! How do you maneuver through the inside of the goal posts? This is where the procedures come into play. By following the instructions established in formal procedures, you can go through the goal posts every time you manufacture. If you properly plan your activities, and follow the instructions given to you, you will increase the likelihood of a trouble-free process. The procedures for maneuvering through the goal posts are established in a set of documents termed *standard operating procedures*.

Standard Operating Procedures (SOPs) These are step-by-step instructions for performing a manufacturing process or operation.

SOPs spell out the steps for a particular work center, such as the Paint section at Everest. They are comprehensive instructions that often contain information regarding safety and training. Step-by-step instructions on how to perform your job, these procedures sometimes include drawings to aid in clarification. A portion of the procedures for painting and curing Everest's bikes follows.

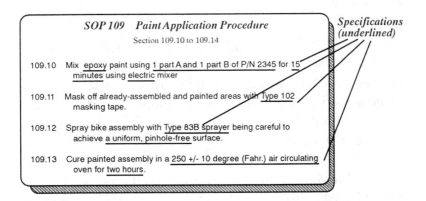

Specifications (underlined)

SOP 109   Paint Application Procedure
Section 109.10 to 109.14

109.10   Mix epoxy paint using 1 part A and 1 part B of P/N 2345 for 15 minutes using electric mixer

109.11   Mask off already-assembled and painted areas with Type 102 masking tape.

109.12   Spray bike assembly with Type 83B Sprayer being careful to achieve a uniform, pinhole-free surface.

109.13   Cure painted assembly in a 250 +/- 10 degree (Fahr.) air circulating oven for two hours.

Notice that good procedures are very precise and often include many specifications. For example, the above procedure contains specification limits for the oven temperature: 250 + / - 10 degrees or in other words 240 to 260 degrees Fahr. By following the standard operating procedures, you can do your job right the first time.

Not following procedures can result in defective product and wasted time, energy, and resources—which means a loss of money. Remember, follow procedures and specifications in order to improve your chances of making it right the first time. Find your procedures, read them, and study them. With attention to the proper manufacturing procedures and specifications, your department and your factory should remain efficient and trouble-free.

TIPS

(1) Mission statements give a company focus and direction. They can also be useful in the various departments and work areas. If your company or work area does not have a Mission Statement, write one and submit it for review.

(2) Know your specifications and procedures thoroughly so that you can make valuable product the first time, every time.

(3) Never pass up an opportunity for more training.

(4) Use the design, materials, and process concept as a problem-solving tool. Try separating these three items in your present operation. If you changed one of the three, would the other two have to change with it? Why?

(5) If you are having trouble making it right the first time in your area, involve your supplier(s) immediately. There is an excellent chance that they can help solve the problem. Keep in mind that you are their customer.

(6) From time to time, you may come across a specification or a procedure that is unclear, incorrect, or can be improved upon. Do not hold this information back. Share it with the appropriate individuals, and get it fixed.

(7) Occasionally, you will follow procedures exactly and still not meet specifications. We will learn some techniques for dealing with this problem later. However, it is extremely difficult to trace a problem in the system if you are not following procedures.

(8) Get to know your management through discussions with your supervisors, managers, and directors. They are there to help you do your job more efficiently.

CHAPTER REVIEW

(1) Ask one of your suppliers for a Mission Statement. Do they have one? If they do, how does it compare to the Everest Mission Statement? Do you feel comfortable that your supplier has the proper quality ethic to supply you with a good product?

(2) Name the titles of two specifications in your own business and give the dates of each. What is the oldest, active, documented specification or tolerance in your business?

(3) Would the part number of a product ordered by a customer be a specification or a procedure?

(4) Is the instruction manual on how to adjust the gears on a bicycle a procedure or a specification? A bolt on the fork of a bicycle is adjusted to a torque value that must be between 10 and 15 ft/lbs. Are these torque values a procedure or a specification?

Chapter Three
Will the Real Customer Please Stand Up!

Rule Three

View your coworkers as your customers, and only pass quality products on to them.

You may not know it, but your most important customers are found in your factory—your coworkers. The coworker who receives your completed operation is your customer. He or she is a source of feedback on how well you have performed your job, and is also very dependent upon your performance—perhaps more than you realize. You are a supplier, and you deliver product to your coworkers. If you pass defective product on to them, they cannot adequately perform their jobs.

Throughout this chapter we will focus on the idea that you are both a supplier and a customer within your factory.

We will also illustrate how you, your coworker, and your factory are part of a larger product team that ultimately supplies the final customer.

TYPES OF INVENTORY

Before we begin a discussion of the customers within a factory, we must lay some groundwork on inventory, the basic material that moves through a factory and eventually is transformed into your final product. Inventory or raw material for most manufacturing processes comes from a supplier who can be either inside or outside a factory. Value is added to the raw material as it flows through your factory and onto your customer.

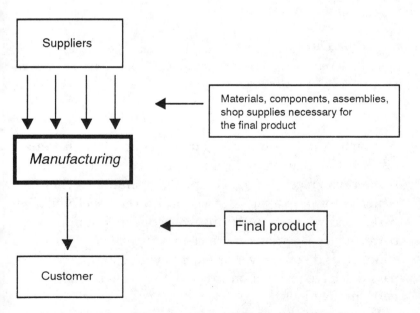

Raw material, work-in-process, and inventory are all synonyms for the material that passes from the supplier to the customer. (A complete description of inventory terms can be found in the Glossary.) This material is valuable and comprises a significant portion of the company's assets. Eventually this material will be converted into money when the product is delivered to the customer.

At Everest Mountain Bikes, Inc., for example, the handlebars, the seat, the tires, the grips, and so on, are all considered inventory before and during the assembly process. Completed bikes that are not yet sold are also considered inventory. The care and development of this inventory is important because it will eventually turn into cash when the bikes are sold to Everest customers.

As a frontline worker or supervisor, it is your responsibility to add value to the inventory as it passes through your area or department. You are considered a supplier to the next operation, and your relationship with the next operation is critical to the overall success of the manufacturing process.

Supplier-Customer Relationship   Concept involving a dynamic working relationship between a supplier and a customer that emphasizes two-way communication.

How do you develop a supplier-customer relationship? One of the best ways is to promote communication among individuals and among organizations. Coworkers need each other to perform their roles well. The next operator needs to receive good product from you, and you need feedback on your performance from the next operator. If this mutual interest is combined with excellent communication, a strong supplier-customer relationship forms. A good relationship based on communication indicates that you are well on your way to becoming an effective team.

NEXT OPERATION AS CUSTOMER

This supplier-customer relationship is encouraged in a variety of factories. The idea originated with Dr. Ishikawa, a leading quality expert, and has continued to grow in usage. Keki Bhote, a senior consultant for the Motorola Corporation, is an acknowledged expert on this subject. He has written several books and articles about the importance of coworkers developing a strong supplier-customer relationship and developing an attitude that the next operation is an actual customer.

Next Operation as Customer (NOAC)   Nontraditional method of manufacturing defines the receiving work station as the *customer*, and the sending work station as the *supplier*, thereby emphasizing the importance of each individual and his or her respective work cells.

Each department can be viewed as a mini-factory that transforms inventory from the previous department into a product, and sends it on to the customer—the next operation.

The diagram in Figure 3.1 illustrates the customer for each department at Everest. Notice the goal posts and the mini-factory within the goal posts for every operation. Each department is responsible for staying within specifications and providing good product for its immediate customer.

Eileen is not the first supplier in the manufacturing process at Everest; the sales department is. However, she is a supplier of wheel assemblies to the Frame Assembly department and a customer of the Bearing Assembly department. In this

FIGURE 3.1

NOAC Method

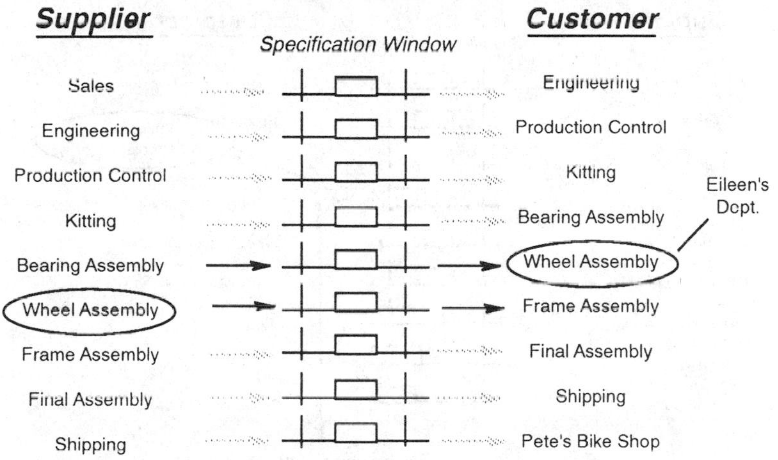

Supplier	Specification Window	Customer	
Sales		Engineering	
Engineering		Production Control	
Production Control		Kitting	Eileen's
Kitting		Bearing Assembly	Dept.
Bearing Assembly →		→ Wheel Assembly	
(Wheel Assembly) →		→ Frame Assembly	
Frame Assembly		Final Assembly	
Final Assembly		Shipping	
Shipping		Pete's Bike Shop	

example, each department is within its specification window manufacturing quality product. Each department respects its job function as a good supplier to the next customer. The mini-factories are centered between the specification limits. Since each department is passing good product on to the next, the process is stable. Product runs through easily, with little wasted time and scrap.

However, let's suppose that the Kitting department of Everest makes a mistake and puts the wrong part number (P/N) bearing into a kit designated for a certain bike. Mistakes can happen. The Kitting department passes the wrong P/N onto its customer—the Bearing Assembly department—that proceeds to assemble the incorrect bearings. Unfortunately, the problem doesn't stop at the next department, it

FIGURE 3.2

NOAC Breakdown

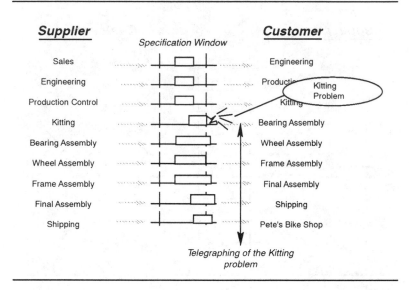

trickles through the succeeding operations. As Figure 3.2 illustrates, the mini-factories move off-center and crash past the goal post limits.

The mistake that occurred in the Kitting department continued through the subsequent departments. Why? Because each department's quality depends on the previous operation. The product from one department becomes the raw material for the next operation. The inventory we discussed earlier is passing through the entire manufacturing process, and if it is disrupted at any operation, it prevents subsequent operations from having a fair chance at succeeding in their activities. This ripple effect is called *telegraphing*:

> **Telegraphing** This is the effect of a change in an early stage of manufacturing that is manifested in subsequent operations.

Telegraphing is a fairly common occurrence in factories—one that can make it difficult to find the original source of a problem. Poor supplier-customer relationships magnify the difficulty and cause confusion. Communication breaks down, and defective product slips through. A department or work center may have trouble in its area without realizing that the cause of the problem actually occurred several steps before. When employees in the Wheel Assembly department started their assembly process, they didn't know that they had received the wrong part number (P/N), making it very unlikely that they could build good product.

In a factory, problems are usually readily apparent. The line shuts down, scrap increases, and questionable inventory builds up. Work piled in front of machines often signals trouble. When problems move outside the factory, customers reject the goods. Frequently, factories have subtle problems that are difficult to understand and track down. A good supplier-customer relationship, based on the Next Operation as Customer philosophy, is a solid, early feedback system that reduces or eliminates problems before they become significant. Effective communication and treating your next operation as your customer aid in reducing defective product and factory disruptions.

MEASUREMENTS—
YARDSTICKS OF PERFORMANCE

Specific tools—called measurements—assist in detecting telegraphing problems, and also help you pass good product to the next customer.

Measurements indicate whether you are within your specification limits or goal posts. Qualities or *characteristics* that are apt to change or vary in size, number, amount, or degree are measured. For example, you may be 68 inches tall, and your father may be 70 inches tall. Your height is a character-istic of you, and your father's height is a characteristic of him. Height is a characteristic that changes or varies, and the change can be measured. This principle is also true of the things that we manufac-ture. There is usually some variation or change between similar materials, and some variation within the same mate-rial over time. In other words, characteristics of materials and processes that are likely to change are measured.

Two categories of characteristics are measured in manufac-turing—*attributes* and *variables*. Each category has its own type of measurement. See Figure 3.3 below.

The term *attribute* is used to describe a characteristic that is either present or missing in a manufacturing product. For example, the reflectors on a bicycle are either present, or they are missing. The quality of the paint job on a bicycle would also be an attribute. If the paint job wasn't exactly perfect, the number of pinholes in the paint job could be counted to provide attribute data, a measure of performance. A common

FIGURE 3.3

Characteristics Measured in Manufacturing

Everest Mountain Bike Measurements			
Characteristic		**Measurement**	
Attribute	An attribute is a quality characteristic of a product that is either present or missing.	**Attribute data**	Paint defects per bike Lost customers due to poor quality New customers due to excellent quality Missing parts per bike
Variable	A variable is any characteristic of a manufacturing process or of a product that may change.	**Variable data**	Length of bicycle chain Tire pressure Weight of bicycle Delivery time

way to use attributes is to count the number of acceptable products manufactured versus the defective product. This is illustrated below in the Everest attribute data for the characteristic of paint quality:

Attribute: Paint Quality

Total Number of Bikes Inspected:	100
Number of Bikes with Paint Pinholes:	2 (Attribute Data)
Yield:	98%

Two bikes with pinholes out of a total of 100 bikes is a numerical count of how well Everest employees did in manufacturing a perfect paint job. Notice the term *Yield*. Yields tell you whether you met your goal (or whether you are within specification). It is a simple ratio of the good parts divided by the total number of parts run through the operation (98 divided by 100 equals a 98 percent yield). Attribute data provide a direct count of your progress toward the prevention of defects.

The other measured characteristic is a variable—a characteristic that can change with time or from one item to another. Some variables are length, width, depth, temperature, weight, speed, customer performance rating, monthly sales dollars, and so on.

Variable: Bicycle Weight

Upper Specification Limit:	29.0 lbs.
Actual Measurement:	28.3 lbs. (Variable Data)
Lower Specification Limit:	27.0 lbs.

The bicycle's weight—the variable data—is 28.3 lbs. Similar bikes won't weigh the same because weight is a variable characteristic. The 28.3 lbs. are well within the specification limits of 27 lbs. and 29 lbs. By measuring variable characteristics, you will know exactly if you are within specification and by how much. With variable data and attribute data, you can determine if you are meeting specifications and passing good product to the next customer. By learning the different uses of variable and attribute data, you can gain full control of the processes within your area and ensure a pleased customer in the next operation.

WILL THE REAL CUSTOMER PLEASE STAND UP!

The real customer—both outside and inside the factory—is a vital part of your existence. Your success depends upon maintaining and improving customer satisfaction by manufacturing a product that meets the customer's expectations and specifications.

The final customer's expectations and requirements must be clearly defined and communicated backwards to each supplier-customer relationship. This communication, called *quality function deployment,* is usually in the form of specifications and process procedures.

> **Quality Function Deployment (QFD)** This quality system is focused on translating the end-customer requirements into measurements for each step of the manufacturing process.

For example, Everest made a mountain bike that was eventually sold to Kevin Miller, a customer who was interested in ruggedness and the ability to free-wheel at low speeds. This can only be achieved if a low-friction bearing is installed at the wheel assembly stage. The wheel assembly requires a light grease instead of a heavy grease on the bearings in order to get the free-wheeling required by the customer. This free-wheeling characteristic can be translated into a set of specifications and then measured during the manufacturing operation. The customer will be present—in a manner of speaking—in every operation of the manufacturing cycle. Specifications and process procedures translate the final customer's wants and needs into every step in the manufacturing process. Measurements of attributes and variables ensure that these wants and needs are met.

The Product Team

With telegraphing in mind, can you imagine what would happen if the inventory brought into the factory at the start of the process was faulty or bad? It would upset the whole process and eventually cause something awful to happen.

Quality at each manufacturing step is the key to success in manufacturing, because each step lays the foundation and provides the material for the next operation. Inventory or raw material that comes into a factory is often passed from two, to three, or maybe more factories in its pathway to your factory.

The importance of this pathway is often overlooked, and few companies take the time to chart the background of the materials that flow through their doors. That's unfortunate since the success of any factory depends on the quality of its raw material. Since this inventory flow involves several organizations and companies, it's important to know the entire Product Team.

You are part of a product team that is much larger than you might imagine. The simple manufacturing diagram in Figure 3.4 suddenly became complex—and we haven't even included a previous generation of companies in the diagram. There are companies that supply Acme Rubber and River Tool and Die. Do you suppose that Kevin Miller, the new owner of a mountain bike, understands all the effort that went

FIGURE 3.4

Everest Mountain Bike Production Team

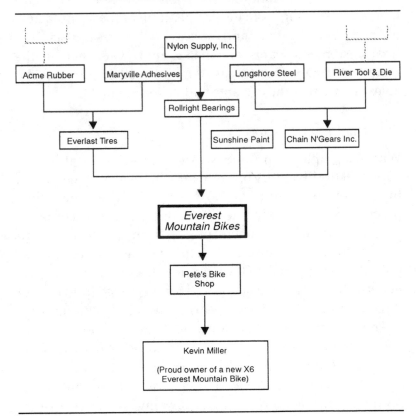

into the manufacturing of that bike, and all the organizations that supported its manufacture? Does Everest Mountain Bike, Inc. know that the total number of individuals involved in its product team is between 700 and 800 employees—or approximately ten times the 74 individuals employed at Everest? Each one of those 700 to 800 employees is contributing to the success or failure of Everest.

Good working relationships with the suppliers on the product team are important. The team will change with time, and it is critical that a written or computerized record be compiled of all activities related to the generation of the inventories. This written record may resemble a family tree.

Genealogy An accounting of the sources of inventory that go into a product, genealogy includes the entire history of how the product came to be manufactured.

The tracking of genealogy means that everything that went into your product is traced—all aspects of the design, the material, and the process in each phase of the manufacturing. For example, in the case of Everest Mountain Bikes, Inc., a complete genealogy will include the process procedures of Nylon Supply, Inc. How do they make their nylon bearings? What temperature did they use? What individuals worked on the nylon? The entire history of the inventory must be compiled, which is no easy task. Fortunately, there is a tool that makes it somewhat easier.

Lot Number This method of identifying manufacturing inventory links a numerical code with the entire history of the inventory in such a manner that all aspects of the operation can be traced. *Synonyms: date code; serial number.*

Items identified by the lot number may include the individual operator who performed the work, the machine where the work was performed, the time the work was performed, the machine conditions when the work was performed, and any components or materials added during the operation. If each supplier in the product team properly tracks all the variables that go into the inventory (or material), then the lot number becomes the link between supplier and supplier, supplier and manufacturer, or supplier and customer.

In the retail world, lot numbers appear on the products you purchase. These lot numbers are usually referred to as serial numbers. With that number, you should be able to trace back and define the entire genealogy of the materials that comprise your retail product.

Kevin Miller received a serial number with his bike. He may never have to use it, but it's there if he needs it. This serial number and the model number of the mountain bike allow Kevin to order replacement parts and even matching paint if needed, from either Pete's Bike Shop or the Everest Factory.

By understanding lot numbers and genealogy, you will understand how your job fits into your factory and how your factory fits into the larger supplier team.

TIPS

(1) Before sending product on to the next operator (customer), ask yourself if you have done everything possible to ensure good product. Are your measurements within specification limits? Be sure to inform the next operator of any potential problems.

(2) Visit your previous coworker (supplier) during the next week, and provide him or her with some information on their performance. Ask how you can be a better customer and communicate your needs better. This technique works well when you identify some positive characteristics of the supplier's work as well as the weaknesses.

(3) Review your raw material for lot number traceability. If needed, could you trace the history of the raw material or subcomponent delivered to your area?

CHAPTER REVIEW

(1) Define the supplier-customer relationship that exists between you and your coworkers.

(2) In a restaurant, is the cook a supplier or a customer to the waitress or waiter?

(3) Would it be good if coworkers occasionally changed jobs? Why or why not?

(4) When baking a pizza at home is your oven temperature a variable or an attribute?

(5) If you are assembling an automobile, and you install a total of 12 fasteners out of a required 12, you have performed a complete job. Is this number a measure of a variable or an attribute?

(6) Notice that in the Everest product team there are two final customers. Why? Is one of these customers also a supplier? Count the number of customers in your factory. How many customers are there after the product leaves your factory?

(7) Lot number genealogy received its start in the pharmaceutical industry. Why and how do you think that happened?

Chapter Four
Spit and Polish

Rule Four

Produce better product and create safe working conditions through attention to detail and cleanliness of your tools, equipment, and work area.

Customers rely on certain sensory clues to determine how well a product is built. They look, listen, and feel to obtain an accurate assessment of their purchase. Quality in the product you manufacture is immediately visible to your customers. Do the parts fit properly? Is the surface free of blemishes, dirt, or imperfections? Does the product rattle during operation? These sensory clues add up. Customers are looking for details that you can introduce by following certain *spit and polish* techniques.

The first step toward achieving spit and polish involves understanding the importance of a clean and orderly factory If the shop is dirty, the product you are building can be contaminated with this dirt and result in poor quality. If the equipment and tools you use in the factory are faulty, misplaced, or poorly maintained, it will be difficult for you to follow manufacturing procedures, and make your product right the first time.

A clean, well-organized shop with well-maintained tools is an important ingredient for manufacturing good product. Even if thorough housekeeping practices are not part of your written procedures, it is your responsibility to develop suitable practices and habits.

A CLEAN AND ORDERLY FACTORY

In most factories, the product flows through the manufacturing work area or the shop floor—an environment that affects the product. This environment includes your work space, the floor, the air, the machinery, and the inventory in the area. If you keep this area clean and well-organized, it greatly increases the likelihood that you will produce good product.

An organized work area is uncluttered; everything has its place. Tools, equipment, and materials are kept in their respective storage spaces. When a tool or a piece of equipment is needed, you can immediately retrieve it, thus saving time and frustration. The common-sense tips on the following page can help you achieve a clean, well-organized factory.

Even the simplest room or work space in a factory can be kept clean, neat, and well-organized. Do not use the excuse that your company does not invest enough money in your area to keep it clean and orderly. You can achieve an efficient

Clean and Orderly Work Space Tips:

- Maintain a sufficient, uncluttered work space.
- Properly maintain your tools and equipment.
- Keep tools and equipment free of excessive dirt, dust, and corrosion.
- Maintain an appropriate storage position for each tool.
- Return tools to their proper position after use.
- Remove unused tools or machinery from the work space.
- Remove unused or scrap inventory from the work space.
- Clean the work space by removing dirt, dust, and excess trim materials.
- Remove excessive paperwork from the work space.
- Emphasize cleanliness in your personal habits.

shop floor work environment by organizing the area, keeping it clean, and removing waste regularly.

You will be amazed at the amount of junk, scrap materials, food containers, and excess paperwork that can build up in a work area. Learn the proper disposal and storage procedures from your supervisor or manager; improper disposal and storage of some shop materials can pose safety and environmental hazards. A cluttered work area makes it difficult to manufacture quality product, and clutter has a tendency to lower worker morale and reduce pride of workmanship. Factory cleanliness is emphasized not just because the factory will look good, but because lack of cleanliness affects so many other areas, including safety.

Customers frequently visit factories. A clean factory increases their confidence that you can produce quality product. You, the frontline worker or supervisor, are a customer representative. Suppliers, coworkers, and customers observe your area when they visit. Use this opportunity to portray yourself positively. After you have removed the clutter from your area, organize your tools, equipment, and materials. You can waste precious time looking for tools that have not been properly stored. When this occurs, you may be tempted to use the "next best thing"—a tool that while similar, was not meant for the job. The use of incorrect tools can result in poor workmanship and reduced quality.

While organizing your tools, verify that they are in proper working order and suitably maintained. Your tools and equipment are a vital part of the manufacturing process, and they need care—lubrication, adjustment, sharpening, calibration, cleaning, and so on, to operate properly. Without this care, the tools will not perform the job you require. Set your standards high. You and your customers deserve equipment and tools that are in good working order.

Before beginning your workday or a new job, take an inventory of the tools you will need. If they are not ready, or they need repair, the correct decision is to stop until the tools are in order. This may take some time, but it is time well spent.

Finally, after you have removed the clutter, returned your tools to their proper storage place, and verified that they are in good working order, it is time to *clean*. Cleaning is vital and part of sound shop practice and discipline. You would not allow your home or apartment to go without cleaning, and you shouldn't allow your work environment to go without cleaning. A dirty work area may contaminate your product.

Dirt is an unwanted raw material that consists of particles originating within the manufacturing process or carried into the shop on clothing and shoes. Fingerprints and hair are also contaminates. Contamination may occur from several sources, including improperly maintained machinery, unclean raw materials, and packaging. Dirt and contamination may also be carried into the work area through the utilities, such as the air conditioning and water supply.

Let's imagine that you are in the Everest Paint department preparing to finish a new set of mountain bikes. If the Paint department is clean, neat, and organized, it will be easy to finish the bikes with a beautiful, glossy finish. If the paint shop is dusty and dirty, there is a good probability that some of the dust and dirt will fall on the newly painted finish and spoil it. Dust and dirt will cause pinholes and dull finishes. The Everest customers will see the poor workmanship. However, if the paint shop is cleaned before the painting process and cleaned after each step that creates dust—such as the sanding step—the probability of a first-class paint job is greatly increased.

Certain manufacturing processes generate their own dirt, such as grinding operations. Filters can be a big help in this battle. They are especially helpful in purifying the water and the air around a manufacturing process.

> **Filtration**   This process involves removing dirt and contamination from air or liquids by passing the air or liquid through a filter media such as fibers or charcoal.

Filters are used extensively in factories to clean the air and to clean processing materials such as water, paints, and chemicals. The proper use of filters can help keep the manufacturing environment clean. Since filters do not have an unlimited capacity, they must be changed periodically. Know the types of filters used in your operation, and change them regularly.

Some manufacturing operations require higher degrees of cleanliness than others. Special uniforms and protective clothing are then used to reduce contamination or increase safety.

Personal Protective Equipment (PPE) Protective clothing or equipment, such as gloves, helmets, shoes, and eye or ear protection, designed to prevent contamination of the product or to protect an individual from a safety hazard.

This equipment is widely used in a variety of forms from the common glove, to the white room protective garments that are often required in clean room facilities. These garments minimize particle or other types of personal contamination from entering the work area.

Additionally, PPE is often required for safety reasons. It protects the wearer from dust, chemicals, or objects that can cause injury in the work environment. Unfortunately, people wear protective clothing more readily when it is designed to protect the product than when it is designed to protect them personally. Protective gear and clothing have been designed to reduce the possibility of serious injury. They are not effective when not worn!

Safety

Although safety is a primary concern in all manufacturing facilities, the most effective safety program begins and ends with you. Most accidents occur from carelessness and lack of attention to the proper procedures associated with machines and equipment. Machines and equipment are especially hazardous when not operated correctly. Operation and maintenance are your responsibility. Following safety procedures and using PPE protect you from job hazards.

Procedures are designed for safe equipment operation. They describe everything from turning the equipment on and off, to performing periodic maintenance on the equipment.

Machines and equipment do not recognize hazardous situations. You must learn the procedures and possible hazards associated with your machinery. A good safety program includes the basic spit and polish principles.

- Use the correct tool for the job.
- Use well-maintained tools and equipment.
- Remove clutter and debris from your work area.
- Wear the appropriate clothing for the job.
- Properly store and handle chemicals.

Learn the correct application for each tool or machine. Use only tools and equipment that are specifically designed for your job function. Maintain your equipment and machines— they are potential hazards when not lubricated or adjusted. Be sure to follow the maintenance schedule, and look for signs of improper maintenance: rattling, leaking lubricant, missing parts, and so on.

Safety procedures apply to more than equipment and machinery; they apply to every aspect of the manufacturing process. If the manufacturing process involves a forklift or a truck, correct traffic procedures should be followed. If the manufacturing process involves chemicals, follow the exact procedures for handling, storage, and disposal. When in doubt regarding a procedure, ask!

If you pay attention to detail, you will greatly reduce the possibility of injury. Cluttered, messy areas in a factory look bad, and they are breeding grounds for accidents. When you become sloppy and careless, you become an accident waiting to happen. Your life and limbs are precious. Take an active interest in your safety.

Common sense is probably your best safety tool. Unplug equipment before you do any type of maintenance. Fully extend ladders on level surfaces, and don't stand on the uppermost steps. Don't assume that vehicle drivers see you. Stand clear of moving machinery and vehicles. Don't handle or dispose of unlabeled chemicals. Don't wear loose clothing around moving machinery parts. Each factory has potential hazards; learn to recognize the hazards in yours. When procedures are followed exactly, most factory environments and operations are safe.

> **Lock Out/Tag Out**   This safety term describes a warning and security system (usually a padlock or a warning tag) used to prevent operation of machinery or electrical equipment by anyone other than the authorized operator. Lock out/tag out may occur, for example, when a maintenance mechanic is repairing a piece of equipment and activation of the equipment during the maintenance procedure would result in serious injury.

Lock out/tag out warns that equipment or machines are out of service. The equipment may be undergoing normal maintenance, or it might be obsolete. If certain locks are in place, do not try to circumvent the locking device. If a warning sign is placed on the equipment that clearly states it is out of service, observe the warning. Serious injuries can occur when a piece of machinery starts operating unexpectedly. Observe lock out/tag out procedures to the letter. If you are in charge of placing locks or tags on equipment, always follow the procedure exactly. Good spit and polish procedures keep the factory—and you—humming.

IF I HAD A HAMMER

Spit and polish requires the proper use and care of manufacturing tools and equipment. Tools are the basic hardware that you use to form, process, or assemble the product during manufacturing. The tools you use are a critical part of your manufacturing expertise. When you use the correct tool, you benefit in several ways. The right tool is safer, faster, and easier to use. To ensure success, use the right tool for the job. One such tool is a *fixture*.

> **Fixture**   This tool is used to hold materials or components while a manufacturing operation takes place.

Imagine a fixture as an extra set of hands that holds the material while it is processed. Fixtures, basic tools in a manufacturing facility, are often used for applications where an attachment to a surface requires accurate placement. Welders use fixtures to achieve accurate placement of their parts.

Another type of basic manufacturing tool is the measuring instrument, sometimes referred to as a gage, which is used to measure characteristics or variables in the manufacturing process.

> **Measuring Instrument**   Measures variable or attribute data. *See gage.*

A measuring instrument provides the numbers and measurements necessary to maintain control of the manufacturing process. It produces the variable data discussed in the last chapter and can be used for attribute data as well. Like a pilot flying an airplane, you monitor the manufacturing process with a variety of instruments. They guide and assist you in accurate manufacturing. Two common types of measuring instruments are *time* and *temperature* instruments. A clock is a time instrument, and a thermometer is a temperature instrument. Clocks and thermometers are essential for successful manufacturing. Good product depends on accurate readings. There are many other types of specialized instruments from light meters, to torque wrenches and vacuum gages. In addition to clocks and thermometers, probably the most practical instruments in manufacturing are the *go/no go gages*.

Go/No Go Gages   These instruments are used to determine if a part is within specification (spec) limits by two consecutive attribute tests—one for the high side of the spec limit and one for the low side.

Go/no go gages usually work in pairs and are most often mechanical gages. For example, two carefully sized pin gages can verify if a 0.125" +/- 0.0005" drill hole is the correct size. If the hole is the right size, a 0.126" pin would not fit into the hole. This is called the *no go gage*, because if the hole is the correct size, the pin will not fit (if it does fit, the hole is too big).

If the no go gage test is passed, then a 0.124" pin is inserted into the hole. This smaller pin is called the *go gage* because a properly drilled hole would accept this pin. These two pins combined together are go/no go gages and can be used in a variety of applications where fast, positive measurements are required. Go/no go gages are easy to use and provide good attribute data. Remember, attribute data represent values that are either present or missing. If a part passes the two-part go/no go gage, then it has a positive attribute. It is likely you will come across a go/no go gage in your work because of its value as an instrument in determining good or bad product quickly and accurately.

On a larger scale, material handling equipment is comprised of the basic tools used to carry and transport other tools and materials.

> **Material Handling Equipment** Machines such as forklifts, conveyors, and robots used to transfer product or material from one manufacturing operation to another are known as material handling equipment.

This equipment transports work-in-process (WIP) and inventory such as raw material and finished goods through the shop. It is the basic mode of transportation of goods through the factory.

Finally, your factory—which creates the environment in which your product is constructed and processed—is a tool! As with a hand tool, the factory should be built for the product that is processed through it. If you are making sophisticated medical products, the factory will be clean, bacteria-free, and supplied with filtered air and filtered water. In most businesses, the factory actually becomes an integral part of the manufacturing process. Semiconductor manufacturers in particular recognize that the facility is an important tool: without the ultraclean environment within a semiconductor facility, microchips would not exist today.

> **Facilities** Category of special tools including buildings, land, and structures that contain and provide utilities to the manufacturing operation and its equipment.

Taking Care of Your Tools

In order to have good, reliable tools, they must be maintained. Manufacturing tools such as process equipment, instruments, facilities, and handling equipment require care. In many factories, the care of machines and tools is provided by a larger system that exists factory-wide. The term for this type of machine maintenance is:

Preventive Maintenance (PM) Work performed on machines on a regular time schedule to ensure that they always run correctly. For example, PM performs lubrication and replacement of key parts on a machine prior to failure.

Proper maintenance of equipment requires thorough planning and technical skills. Companies often involve experts in machine maintenance and enlist the suppliers of the tools and machines to develop a preventive maintenance program. A good maintenance program reduces problems and/or unscheduled breakdowns. You may be called upon to do some of the maintenance on your equipment, or you may be an inspector of the work performed. In any case, learn as much as you can about the preventive maintenance program in your company.

With certain pieces of equipment and all instruments in the factory, a periodic determination of accuracy is essential.

Calibration Determines the accuracy of an instrument by using a reference standard and adjusting the instrument with the appropriate correction factors, if required.

Calibration is a distinct class of preventive maintenance. Special skills are required since the instruments used in factories often are sophisticated and require specialized support equipment to properly perform the calibration. If an instrument is out of calibration and providing wrong variable or attribute data, it could virtually throw the entire manufacturing process off balance. Decisions are made based on the readings of these instruments, so they must be extremely accurate. Verify in your own area that all of your instruments possess an up-to-date calibration label, and that they are accurate. If you are responsible for periodic checks of your instruments, perform them promptly and carefully.

In many factories, a Quality Assurance department is responsible for the calibration of most instruments. These calibrations take place on a regular schedule by experienced, qualified individuals. If you suspect that any of the instruments in your area are out of calibration, *immediately* notify your supervisor and the Quality Assurance department to have the gage recalibrated.

THE LAST DETAIL

A clean, organized factory where everyone uses the correct, properly maintained tool for the job helps ensure quality product. However, something is still missing. . . *The Last Detail*. In the past, products were sometimes manufactured with parts that did not fit together correctly, or the finish was less than perfect. Products were not necessarily rejected for excessive rattles or defects. Some imperfection was considered acceptable. With increased competition, customers are less likely to buy products with obvious imperfections.

If customers can't find quality in one product, they look elsewhere. Today's customers understand world class quality and are demanding it. They expect that the parts in the product they buy will fit precisely, and the finish will be flawless. In order to be competitive, manufacturers must be detail oriented.

Observe your product as your customer would. Listen to the product if it has moving parts. Is it quiet and smooth running? Feel the product after you have finished it. Does it have the surface texture your customer is expecting? Would you pay money for this product? If you can't answer "yes," your work isn't done.

> **Cosmetics** The appearance of the product, especially as it relates to the final finish, paint, color, and or workmanship, is known as *cosmetics*.

Cosmetics are those characteristics that can be seen. Most of us make judgments based on appearance. Part of being a world class manufacturer involves understanding that cosmetics are important. If customers see defects in a product, they will assume that there are defects in areas of the product that can't be seen. The customer is probably correct; if you don't get the cosmetics right, you haven't finished the job.

Attention to the details of cleanliness and organization in your work area will help you attend to the detail in your product. These are good habits that will make you a world class manufacturer.

TIPS

(1) Get rid of items in your work area that are not useful, and organize the rest. A clean, orderly factory eliminates waste and makes troubleshooting easier. The work space tips provided earlier in this chapter are a good daily checklist.

(2) Never trust your instruments completely. This will be discussed in later chapters, but remember to challenge the data you are receiving from your instruments on a daily basis. Rule of thumb: instruments should be 10 times as accurate as what you are trying to measure.

(3) Visually scan the corners, edges, and surfaces of your product. Think of ways to improve the DETAIL of the product.

(4) Cosmetics are the customer's first impression of a product's quality. Make a good first impression, but make sure that this beauty is more than skin deep.

(5) Safe work habits begin with a good attitude. Approach your job with a positive, attentive attitude. Focus directly on your job, and avoid daydreaming and other mental activities that will prevent your common sense from operating.

(6) Personal protective equipment frequently is not worn. Don't get caught in that trap. Enjoy a long, healthy career. Always wear appropriate protective equipment.

(7) Daily personal exercise prevents accidents. This is due in part to the muscle relaxation that occurs after exercise. Consider a reasonable exercise program that fits your health requirements.

CHAPTER REVIEW

(1) When you are shopping for a product, and there are two similar products at about the same price, do you look for the *details* that might differentiate the products, such as how the parts fit and the surface finish? What do the cosmetics tell you about the product?

(2) Look up HEPA in the Glossary. Could this type of filter improve your manufacturing facility?

(3) A human hair is about 0.004" in diameter and can be a major source of particle contamination. Does your

factory understand this source of contamination in its critical areas? What kind of protection is used?

(4) Could a pair of pliers be considered a fixture? Name two fixtures used in your factory.

(5) Predictive maintenance is a type of preventive maintenance. Look up the definition, and determine if your factory has predictive maintenance.

(6) Are the air conditioners and air conditioning filters in your factory part of the manufacturing process? Explain.

(7) After checking the terms in the Glossary, explain the difference between gage repeatability and gage reproducibility.

(8) Would lock out/tag out be necessary in an automobile repair facility? Explain.

Chapter Five
Stop the Presses!

Rule Five

Communicate problems clearly to other members of the manufacturing team, and learn to shut down the process responsibly.

D id you ever have a day when nothing seemed to go right; or a day when you knew something was wrong, but you couldn't put your finger on the problem? You are certainly not alone. In manufacturing plants, unexpected or unplanned things happen. This is a complex world: people make mistakes; machines break down; unpredictable events can and will occur. Responding quickly to a problem requires skill and expertise—a fast response minimizes the loss to the company and prevents large disruptions in the operation. This chapter will review some of the skills needed when problems arise in the manufacturing process.

After a problem develops, it is only a matter of time before the consequences are felt in the business. Therefore it is very important that the problem be detected early and treated early. This is similar to personal medical problems, since early detection and correction can minimize damage to the bodily systems. Lack of action and treatment allow the problem to continue, resulting in serious consequences. In manufacturing you have an advantage over living systems. You can STOP the motion—freeze the action—until you correct the problem. You can then continue manufacturing after you have fixed the problem. Continuing production without repairing the problem has staggering consequences—bad product, lost time, rework, and repair costs.

COMMUNICATE THE PROBLEM

When in doubt, communicate the problem to the rest of your manufacturing team and your supervisor or manager. If the problem warrants further action, stop the process in a responsible manner and study the situation. Everyone in the manufacturing facility is responsible for the quality of the product. A problem in manufacturing should be communicated throughout the factory and appropriate action taken. If all employees are standing guard and prepared to stop the action when a problem occurs, the chances of manufacturing bad product will be greatly reduced. When a problem arises, the best approach is the team approach. Consult frequently with your supervisor and coworkers when you suspect a problem. Since everyone in the factory will be affected by the problem, most will be more than willing to help with the solution. In a factory setting what's important isn't *who* finds the solution, but that a solution is found.

Be sure to remember that certain types of manufacturing processes and equipment cannot easily be shut down without risking further damage. It is important to know your process well enough to stop it safely, if required, without risking further product problems or potential safety problems.

No Brainers

How do you know when to stop the action? In many cases, it is relatively clear-cut. There are five main types of problems that frequently occur:

(1) *Machine Problem*   When your machine or equipment malfunctions, or when it starts to make unusual noises or leak fluids; it should be safely shut down until it can be repaired or maintained.

(2) *Material Supply Problem*   If you have not received sufficient parts from your coworker, your supplier, or the next operation has received too many parts, it is normal to shut down your operation.

(3) *Faulty or Poor Quality Material Problem*   When you receive faulty parts or poor quality materials from a supplier or from a previous operation, you should discontinue manufacturing until the supplier is notified and the problem corrected.

(4) *Out of Tolerance Process or Product*   If your product or your process does not meet specification and is considered *out of tolerance*, the operation should be stopped at once.

(5) *Safety Problem*   If the manufacturing process or the product being manufactured are unsafe, the

operation should be discontinued until the problem has been resolved.

The preceding problems, which may develop without notice in a manufacturing process, are sufficient reasons to stop the process. When these conditions occur, shutting the manufacturing process down quickly minimizes further problems. Fortunately, there are several ways to reduce the probability of a shutdown situation. Proper maintenance of the equipment and process control are key elements in maintaining an operation without disruptions. Two types of control methods—*statistical process control (SPC)* and *precontrol*—are used as early warning indicators of problems.

Statistical process control (SPC)—Involves controlling and improving a process by recording sample readings of the process, analyzing the readings with statistics, and performing process corrective actions when the process starts to drift out of control.

Precontrol An early warning system for shutting down the manufacturing process, precontrol does not require an understanding of mathematics, but rather depends on an operator recording data points in one of three specification zones—the green zone for acceptable, the yellow zone for caution, and the red zone for unacceptable. *See the Glossary for more information.*

Both SPC and precontrol signal the need for corrective action in a manufacturing process. More and more companies are beginning to use these systems for early detection of problems. They are valuable tools, and you should make an active effort to learn as much as possible about one or both of the systems. With SPC and/or precontrol, you can discover trends and correct them before they cost the company time and money. The basics of statistical process control will be covered in depth in a later chapter. Early warning systems are no substitute for common sense. Stop the process when you think there is something significantly wrong. Discuss the problem with your supervisor or plant manager. Use your fellow workers as a team to solve the problem.

THE RED FLAG

Sometimes you will actually get a sense that things aren't quite right *before* problems occur in the manufacturing process. You are actually getting clues that something is wrong, but you either don't know how to read them, or choose not to. These early clues that a problem is developing are *red flags*— feelings or thoughts that something is not quite right. It is not unusual to dismiss or overlook these feelings until something erupts into a major problem. Then, you will look back and remember that you had noted it as a potential problem.

> **Red Flag**   This is an indication of a potential problem that should be examined before a major problem disrupts the process or product.

The following is an example of a red flag situation: In the Everest Paint department, bikes were painted blue on Tuesday, and others were scheduled for red paint on Wednesday. Jaime, the operator responsible for moving the bicycles from the Paint department to the Frame Assembly department, noticed on Tuesday that the eleventh bicycle off the line that day was not actually blue in color—it was *purple*. Since 100 bicycles were painted that day, Jaime reasoned that it was probably a mix-up in the paperwork. Jaime did what had been done in the past: He placed the purple bike in the warehouse (where mistakes often find their home) until new paperwork for a purple bike came through.

Did Jaime make a mistake? Yes! He should have questioned the purple paint, and Paint department activities should have been halted until the cause of the purple bicycle was discovered.

Later, it was discovered that the purple bike was not simply a case of wrong paperwork; the purple color was a red flag indicating a process out of control. In fact, the red paint mixing vessel had sprung a leak and was spilling into the blue paint vessel creating the purple color. Unfortunately, an additional 50 bicycles were painted purple before the problem was corrected. A major problem could have been avoided with a few simple questions and an investigation. Be alert— ask questions!

FISHBONE DIAGRAMS

Stopping a bad process is just the beginning of a skillful approach to a problem. After stopping the process, it is important to quickly resolve the problem in order to continue manufacturing. When you thoroughly understand your process, you will know most of the possible causes of the problems. A *fishbone diagram* is a visual method of displaying all possible causes of a problem. When a problem arises, you can look up the most probable cause on the diagram, and proceed with the corrective action.

> **Fishbone Diagram**   A cause and effect tool developed
> by Dr. Kaoru Ishikawa, an authority on quality control
> in Japan. The fishbone diagram is a graphic representa-
> tion of all possible causes of a particular problem. Each
> spur on the skeleton indicates a possible cause. The head
> of the skeleton represents the problem.

Fishbone diagrams are powerful tools that can be used to
determine the cause of a problem. Potential problems and
their possible causes should be noted on a fishbone diagram
prior to the start of the manufacturing process. This takes
some time and thought, but it is worth the effort. You can see
in the fishbone diagram in Figure 5.1 six possible reasons for
a rear wheel imbalance problem at Everest. When an imbalance situation occurs at Everest, the operators can shut down
the process to prevent bad product. Then, each of the six
potential causes of the imbalance can be investigated until the
actual cause is discovered.

Trust your instincts, and use your common sense. If your
machine is malfunctioning, or you are manufacturing poor

FIGURE 5.1

Fishbone Diagram

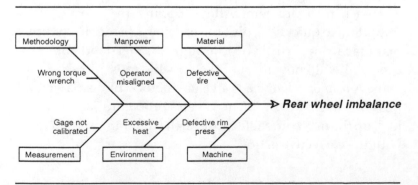

quality product, *stop* manufacturing, and check your fishbone diagram for possible causes of the problem. Discuss the problem with your supervisor and coworkers. If you feel that something may be wrong, but you are unsure, stop and investigate.

TIPS

(1) It is less costly to shut down a faulty operation than to run bad product. Avoid the temptation to continue manufacturing when the process or your machinery goes out of specification. Communicate the problem quickly to spur your manufacturing team to action. Stop the manufacturing process if necessary, and fix the problem.

(2) Trust your instincts. If something doesn't look or feel right in the manufacturing process, question it.

(3) A drop in process yields is an indication of a problem in your process that should be investigated.

(4) Shutting a process down can be unsafe and disruptive to a factory. Rule five advises that the process be shut down in a "responsible manner." Do not act alone in these situations—inform your supervisor, management, and/or coworkers about the problem prior to shutting down.

(5) When problem solving, take a broad view of the problem. Analyze the main ingredients in manufacturing—design, material, and process. Plug in the fishbone diagram.

CHAPTER REVIEW

(1) When a serious quality problem occurs with a supplier, do you believe the supplier should be disquali-
fied until the problem is corrected? Explain your
answer.

(2) If all the employees in your organization committed
themselves to shutting down the process when a
significant problem occurred, and fixing the problem immediately, would your daily routine change?
How many times a day, a month, a year would the
manufacturing process be shut down? Would this
number change upward or downward with time?
Explain your answer.

Chapter Six

Waste Not/ Want Not

Rule Six

Keep good product moving by minimizing wasted time, energy, and materials.

W alk around your factory, and observe a few details. Do some of your coworkers have less work than others? Are there any areas in your factory where the work appears to be piling up and unequally balanced? It would be surprising if every area in your factory were perfectly balanced, and everyone had the same workload. Even in the best of factories, differences in workloads often exist.

An uncomfortable pressure exists in factories with imbalanced work areas. In this chapter we will discuss techniques that can turn this problem situation into an opportunity. In the last chapter we concentrated on stopping the process when defective product appeared. Now we will concentrate on keeping good product moving.

Good product can keep moving only if you identify all the factors that slow the process down. Factors that can potentially slow the flow of good product are almost limitless. Lack of adequate training, faulty design, faulty raw material, faulty process, employee absenteeism, lack of team cooperation, and poor maintenance are some of the common factors. If one or more of these factors is affecting an operation, the flow of good product will be stopped or slowed down.

This results in a bottleneck situation similar to the one found on a freeway after an accident. Cars back up when there's been an accident; they move very slowly or not at all. Beyond the accident, lanes are open and free flowing for the cars that have managed to pass the bottleneck. The goal in the factory is to keep all lanes open and the product flowing. Each operation and each worker are dependent on all the other operations and workers.

As stated earlier, you play a critical role in successful manufacturing. If you haven't received adequate training, for example, or you are frequently absent or tardy, the product flow will tend to bottleneck in your area. A bottleneck in your area will make you feel stressed and pressured. The workers in the operations after yours will be standing around waiting for product.

You are an important part of the team. Imagine a baseball game with the bases loaded. The player who would have been the fourth batter didn't show up for the game. The players on the bases, the pitcher, and all the other field players stand around waiting. The game can't go on until a replacement is found. The previous strikes and home runs don't count unless the game can go on. A bottleneck in the manufacturing process is a similar situation. Raw material can't become finished product without everyone doing their part. The team is only as strong as its weakest player. Employee weakness is only one possible cause for a bottleneck of product. But you have the most control over this area. If product builds up in your area because you are having trouble following the procedure, ask for help. Maybe the procedures need to be rewritten so that they are easier to follow, or maybe you need additional tools or training. Finding the solution to the problem is much easier than struggling with the stress and pressure of a never-ending pile of backed-up work.

Everest experienced a work backup problem. Product was building up in the Paint department, and Rick was falling behind because the nozzle on his sprayer kept clogging up. He tried other nozzles, but they had the same problem. Jaime, the next operator, wasn't getting painted frames. He began to wonder why, and asked Rick if he could help. Rick was a little grumpy as a result of all the problems he was having, but he explained the situation. Both Rick and Jaime decided to ask the supervisor if she had any ideas. They decided as a team to analyze the design, material, and process to locate the problem. They soon discovered that the design and procedures required that a certain type of paint be used with a certain nozzle size. They also discovered that the nozzle size was too small for the thickness of the paint. In order to keep good product flowing through their area, the procedures needed to be rewritten. A larger nozzle is required for the job. Rick probably would have discovered this by himself eventually, but the team solved the problem much faster.

Earlier you saw that a problem situation or bottleneck could be turned into an opportunity. Discovering that a worker needs training, or that procedures need to be rewritten, are opportunities to improve the process. The problems point to areas that need improvement. When the problems are solved, the processes will be stronger and better. Hopefully, if you keep notes and modify your processes, you won't have to solve the same problem twice.

WASTE NOT/WANT NOT

World resources are limited. This is especially true in manufacturing where we don't have an unlimited amount of time, materials, or energy at our disposal. We must conserve resources and prevent bottlenecks, scrap, and other disruptions from consuming our resources. Our resourcefulness is measured by the amount of waste we generate—if we generate small amounts of waste, we are resourceful. *Waste* defines the time, materials, and energy not turned into valuable product. Recognizing this waste is an important skill. Manufacturing plants are prone to high amounts of waste because one problem leads to another. Instead of adding value to raw materials, waste becomes a financial burden.

Your company is paid for the final product and the recyclable materials—they generate wealth for the factory. However, no one wants to pay directly for the waste! Waste of valuable resources—including time—is a loss, pure and simple. Total waste in many United States factories is in the 30 percent to 40 percent range. The opportunity for improvement is quite significant.

There are many ways that you can reduce waste. The obvious ones include turning off lights, machines, and tools when not in use; monitoring the amount of water used; and monitoring the amount of paper used and reused. Waste can also be reduced by making the product right the first time, keeping your factory and work area clean, and using the right tool for the job. In other words, most of the techniques described in this book will contribute to the reduction of waste and the promotion of resourcefulness. By applying these *smart* techniques, you can make a significant difference. The ultimate goal, of course, is to achieve zero waste, but that is not always realistic. It is more sensible to set a target and consistently work toward the target every day. By building quality product through attention to detail, and controlling your manufacturing process, you can reduce waste. A small improvement every day makes practical sense.

JUST-IN-TIME

There is a system called *Just-In-Time (JIT)* that helps factories to remove some of the waste. JIT systems have an underlying principle—they prevent excess material from building up inside the factory by controlling the product flow from operation to operation. Work-in-process does not move to the next station until that station can process the inventory within a reasonable period of time. This method of controlling inventory flow is accomplished with a *signaling* system, which is similar to a traffic cop. JIT assures that product—or work-in-process—moves to the next operation only when it is needed, and the next operation is prepared to receive the material.

> **Just-In-Time (JIT)** Reduces the total inventory in a factory by delivering material to the next operation only when the next operation is prepared to process the material. JIT is accomplished through accurate scheduling or signaling systems from the next operation. *Other terms are stockless production, kanban, synchronous manufacturing, pull system.*

Some JIT *signals* are generated by computers that schedule and predict the amount of material needed for an operation.

Another JIT system is more basic and relies on visual cues. When a worker notices that the worker in the next operation is running low on inventory, additional inventory is sent. The JIT signal is the visual cue that the next worker needs more material. One of the simplest JIT systems sends product to the next operation in a reusable container. When the container is empty, the frontline worker sends the container back to the previous operation for more product. The empty container is a JIT signal. Since good product is moved to the next station only when needed, excess stacks of material are reduced. Regardless of the type of signal used, the flow of materials from station to station is performed in the following manner.

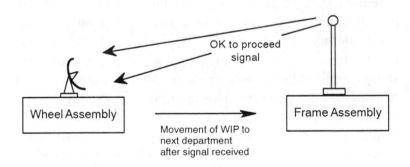

OK to proceed signal

Wheel Assembly

Frame Assembly

Movement of WIP to next department after signal received

Many factories have a tendency to become clogged with excess materials and inventories due to bottlenecks and poor planning. It is common to observe piles, boxes, carts, and totes scattered throughout the factory with product that is not moving. It is waiting for the next operation. This excess material severely limits the efficiency of the plant.

Quality problems can lurk undetected in these stacks of material. The stacks can also hide underlying problems within departments such as poor workmanship or poor machine maintenance. The JIT system reduces the amount of work in any one area. Since signals control movements of material, a problem at any one operation will shut down the entire manufacturing line.

When a problem occurs, the signal will not be sent to the previous operation, and that operation will not signal its previous operation, and so on—back to the beginning of the manufacturing line. Operations will come to a screeching halt. When this occurs, the problem will be visible and not masked in the piles of material throughout the factory. JIT systems surface problem areas quickly. They work most effectively when the following foundations are in place.

Just-In-Time Foundations

(1) Straight Pathway of Flow

(2) Line Balancing

(3) Quick Tool Changes

An understanding of these foundations will help you streamline your department or work station. If you already have JIT in your factory, you will have a better understanding of the system. Let's examine how these techniques help reduce wasted time, space, and material.

The Straight Pathway

You probably learned in school that the shortest distance between two points is a straight line. If you can travel in a straight line to a destination, it will take you less time than a curving, turning route. Companies want to remain competitive, and many are trying to implement this straight line concept. If product moves through the factory in a straight line, it will save both time and money.

Frequently, manufacturing plants look more like race-courses than efficient factories. Sometimes this has to do with available space, or sometimes a company has grown without an efficient plan for product flow. Everest uncovered this racecourse pattern when reviewing the flow of product through the factory.

As the mountain bikes proceeded from one operation to the next, they turned, twisted, and moved from one room to another. The product traveled 1,280 feet in its manufacturing route, but this product flow didn't show the whole story. At each operation, Everest added key raw materials or components to the mountain bikes such as tires and handlebars. These raw materials were stored at various places within the factory and moved when needed. When they analyzed the movements of the materials that fed the assembly line, they saw the same twisting, turning pattern, and a lot more wasted movement.

When the movement of the raw materials was combined with the work-in-process, a total movement of 16,000 feet was

FIGURE 6.1

Original Product Flow

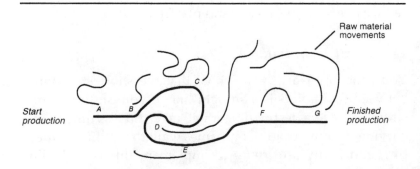

Manufacturing Route 16,000 Feet

discovered! See Figure 6.1.This movement takes time and energy. It also masks other problems that need attention, since most of the time is spent moving material from one location to another.

With planning and some manufacturing skill, Everest saved time and space by reducing the distance the materials traveled through the plant. Everest implemented a *flow through factory* (see Figure 6.2 below) approach, and aligned the operations within the factory. Everest moved assembly operations and tore down some walls. Conveyors were installed, and storage containers for raw materials were strategically placed. The end result was dramatic.

Everest reduced its manufacturing route from 16,000 feet to 2,000 feet! Straight lines were not achieved everywhere, but efficiency was increased. Some U-turns and 90 degree turns were installed because the original building was not built as a long rectangle. Much of the wasted motion in the facility was removed, and it was easier to locate problems and move materials. However work-in-process still had a tendency to pile up in certain areas. Line balancing was needed.

FIGURE 6.2

Flow Through Factory Approach

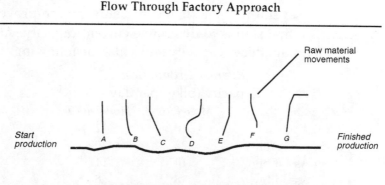

Manufacturing Route 2,000 Feet

Line Balancing

Manufacturing a final product consists of multiple steps or operations. This is part of an overall journey from raw materials to the final product. During this journey, the raw materials go through a variety of phases, including processing, converting, assembling, and anything else involved in a particular business.

Each step is not exactly the same; some will require more time than others to complete. During some operations, multiple steps are completed. When the manufacturing of a final product requires numerous operations, bottlenecks are almost guaranteed. This is due to the fact that most businesses cannot afford unlimited equipment or unlimited workers to build product. There will always be some operations that are slower or faster than others.

After a factory implements JIT systems, it must balance the factory in terms of the capability at each step. Each operation defines its capability in terms of *capacity*.

Capacity   The maximum output of a machine, a work center, or a factory.

There are several ways to define, measure, and improve capacity. The first step is to determine current capacity. At Everest Mountain Bikes, capacity looks like the following:

Everest Production
Capacity (bikes per day)

Operation	Model X6	Model 400T
Kitting Dept.	10	12
Bearing Assembly	24	20
Wheel Assembly	14	11
Frame Assembly	8	8
Final Assembly	24	24
Shipping Assembly	50	50

As you can see, there are a number of differences in the bikes that can be produced each day in the various key departments. Look at the table. How many Model X6 bikes can Everest produce daily? How many 400T bikes can be produced daily? Suppose there was a mix of both models? The answer is simpler than it looks. The real capacity of the entire plant is only *eight* bikes daily. It doesn't matter that some departments have the capacity for up to 50 bikes a day. Remember the *signal to proceed*? In the example above, the Frame Assembly department is sending out only eight signals a day to the previous departments. The Frame Assembly department is a bottleneck since it is the department with the smallest capacity. Therefore, Frame Assembly becomes the pace car that all other departments naturally follow. Good manufacturing skills emphasize that the true capacity of a factory is set by the department with the smallest capacity.

There are many ways to balance a factory to achieve a more uniform capacity throughout the shop. Most of the line balancing techniques are straightforward and require nothing more than common sense to find the best way. Additional personnel or equipment, for example, may improve the capacity of the area. Many companies elect to work overtime in the bottleneck areas. In the case of Everest, there is a very high probability that the Frame Assembly department will be required to work overtime when Production Control schedules more than eight bikes a day.

Overtime is a short-term solution. The real solution to the problem (assuming Everest wants to make more than eight bikes a day) is to increase the capacity of the area. At Everest, there is only one person who can balance a frame once it is assembled. She can only balance eight bikes a day. If another worker were trained to balance bikes, the plant capacity would increase to 16 bikes daily, and would probably solve some scheduling problems for the factory.

Now we have product moving in straight pathways, and we have identified capacity weaknesses and corrected them. If we can perform quick tool changes, our factory will be humming efficiently.

Quick Tool Changes

Flexibility is the king of competitive manufacturing. The ideal manufacturing system runs many different types of product through the same equipment. When Henry Ford manufactured his first cars in an assembly line fashion, he didn't care whether the equipment could run a variety of parts through, or how quickly the equipment could be changed to accommodate other parts. The reason for this was simple: although he was very innovative in some of his other approaches to manufacturing, Henry Ford didn't recognize the importance of flexibility in manufacturing. Ford's production line contained many tools made specifically for a certain model car. These tools were used for one model and then thrown away or reworked for new designs.

Today, many of the tools used in manufacturing are worth millions of dollars. Consumers are more selective—they want specific features or products that require the manufacturer to produce a variety of products without extra cost. The manufacture of a variety of products with low costs requires flexibility in factories. Most modern tooling systems allow some flexibility, but tool changes take time, and time is money. The faster you can change the tools, the lower the cost, and the more likely it is that you will remain competitive. Many newer types of manufacturing equipment are programmable by computers and allow almost instantaneous changes to the machine for different part numbers (P/Ns) and designs.

Some companies can't afford sophisticated computer controlled equipment. Fast tool changeovers can be achieved by doing tool changes off the assembly line. In other words, plan before putting the tool into production. If your tool changes are prepared in advance, valuable manufacturing time will be saved. This simple step reduces the likelihood of a bottleneck in your area.

For example, Rick in the Paint department at Everest planned for quick tool changeovers. The paint sprayer is a *tool*. Rick and Jaime discovered as a team that a larger nozzle size improved the spraying of a particular type of paint. It takes Rick approximately 17 minutes to change a nozzle and adjust it for the various paints. This doesn't sound like a lot, unless Rick is required to make several changes during the day to accommodate different paints. If he makes three changes in one day, almost an hour will be spent changing and adjusting nozzles.

How can Rick improve the output of his area by achieving faster tool changes? The simplest method would be to have a spare sprayer—one with a smaller nozzle—and a second sprayer with a larger nozzle. Each sprayer could be readjusted to the different types of paints. Rick tried this technique and reduced his changeover time from 17 minutes to 2 minutes, greatly increasing the capacity of his department.

There are many different methods of improving tool changeover time, depending on the type of operation. This is an area where you can use some creativity, and suggest different methods to your coworkers, supervisor, or manager. Quick disconnects may work, where tedious traditional fasteners slow the tool changing process. Again, use the team approach, and the improvement will be more substantial throughout the plant. Eventually, your team will begin to see the big picture of how each individual contributes to the smooth flow of product through the factory.

CHECK IT OUT

Before beginning your workday, take some time to anticipate things that could go wrong; make a checklist of these possibilities. Then go through the list of items, and take the necessary steps to ensure that these things *don't* go wrong. For example, *machine breakdown* might be on the list. Go through all the items on the fishbone diagram to ensure that everything on the machine is in good working order. *Defective raw material* might be on the list. Check it out. Don't go through the whole day using defective material.

Your checklist will probably contain quite a few items. It will take some time to ensure that these things don't go wrong, but it is time well spent. Good product won't be moving through your area while you are performing your preventive maintenance, but this procedure helps ensure that the product will keep moving throughout the day. At first you probably won't be able to anticipate some of the things that can go wrong. Add the things that do go wrong to your checklist; ensure that they don't go wrong tomorrow. If you learn from your mistakes and keep records of them, you become a skilled, valuable employee. If you are prone to making a certain mistake, you can almost bet that others are prone to making the same mistake. When you gain the knowledge that prevents the mistake, the whole company can benefit.

Planning goes hand in hand with the checklist; it keeps good product moving and reduces the likelihood of problems. Good planning involves preparation. Prepare your tools and materials at the beginning of the day.

You don't have to blunder through the woods waiting for things to pop up. You can think, analyze, and anticipate what you might find in the woods, and develop a plan to turn it to your advantage. Additional planning techniques will be discussed in later chapters.

TIPS

(1) Keep a planning checklist for your operation. Include items such as tool preparation, maintenance, a fishbone diagram, and so on.

(2) Think of JIT manufacturing as a chute. Once the process is started, quality product easily rolls down the chute and exits the Shipping department with minimal effort.

(3) Learn to watch for *signals to proceed* in your factory, and respond accordingly.

(4) Planning ahead is only valuable if the plan is followed. In your operation, your *plan* is provided in the *bill of material* and the *standard routing*. Look these terms up in the Glossary.

(5) Before each work week, list all the events that could possibly go wrong in your area: maintenance problems, poor workmanship, lack of cleanliness, and so on. Then ensure that each potential problem is prevented. If a problem appears unavoidable, gather your team together, and figure out a solution.

(6) Look up the term *cycle time* in the Glossary. Cycle time is a tool you and your team members can utilize to measure your progress as you keep good product moving.

CHAPTER REVIEW

(1) Do you think that numerous manufacturing steps increase the physical distance the product moves? Is there an advantage to fewer steps? If so, what are the advantages?

(2) Draw a diagram of how the work-in-process moves through your manufacturing area. Is the flow simple and straight? If not, what suggestions do you have for improving the flow?

(3) If there are two bottlenecks in a plant, and they both have the same capacity, which of the bottlenecks do you believe is more important, the first one or the second one? Explain your answer.

(4) What is the average cycle time of your operation? Give two possible ideas for reducing cycle time.

Chapter Seven
Keep It Simple

Rule Seven
Keep the design, process, and material simple.

S ome of you may remember the large radios that people used to listen to in the 1930s and 1940s. Perhaps you also remember the big, bulky televisions with small screens from the 1950s. Consider the size of cars in the 1950s and 1960s. And when computers first came into use, *one* computer filled a large room. What happened to all of these huge products; why don't we make them anymore? *Big* in manufacturing has gone the way of the dinosaur—and probably for some of the same reasons. They were simply too big and bulky to support on our earth's limited resources.

Today we have radios and TVs that fit into shirt pockets. Computers are the size of small briefcases and shrinking

85

rapidly. Small cars dot our highways. People love the smaller, leaner, lighter products that have fewer parts and require fewer raw materials. The factories that produce these goods also run on the smaller, leaner, lighter concept. The dinosaurs of yesterday are extinct, and you have to learn the new ways to stay in the ball game.

Keep it simple. These three words convey a powerful concept that all employees in a factory can benefit from. Designs that are too complicated result in wasted materials and extensive manufacturing problems. Manufacturing processes with too many steps are prone to bottlenecks, lose valuable time, and increase the likelihood of defective product. This chapter will address some techniques that are useful in simplifying the design of both the product and the manufacturing process.

In your present position as a frontline worker or supervisor, you may not be involved in the design of the product or the process, but more and more companies are including the frontline worker in the design process. When you are asked to participate in the design process, you will know exactly what to suggest and improve upon.

SIMPLICITY OF DESIGN

Design is an integral part of the manufacturing world. It defines how the final product will look and function. In previous chapters design was introduced as one of three key elements that combine to form a product: the design, the process, and the material. If the design is well thought out and simple, final product can be manufactured with high yields and a relatively simple manufacturing process. However, if the design is too complicated, a new form of waste will enter the equation—design waste. This type of waste will last forever, or at least until the design is changed. Not only will the materials cost more for the design, but at times the manufacturing process will be needlessly complicated and excessive. A poor design costs money and takes many forms.

- Excess weight
- Extra parts
- Excessive assembly time
- Unreliable product

- Excess space
- Specialized, costly tools
- Chance of incorrect assembly
- Unsafe product

A good design takes advantage of every ounce of weight and every inch of space. It takes time and skill to develop. Many industries have difficulty developing efficient designs that are compatible with the manufacturing process. Most of the recent improvements in design technology have come from the auto industry. With a part count that can exceed 10,000 components, the automobile is a prime example of a product that requires a good, simple design. The automobile industry has provided us with three specific principles:

- Lean design
- Lock and key design
- Value analysis of the process

These techniques form the basic foundation of a resourceful product and process design. Because these principles relate directly to the *producibility* of the product and elimination of waste, they are gaining ground in manufacturing circles and educational forums beyond the automobile industry—similar to the way JIT gained ground several years back. Recently, many manufacturers have become committed to designing their products with manufacturing in mind.

They've learned that the interrelationship of the design, the process, and the material is necessary to produce a worthwhile final product.

By learning the basics of these three design principles, product designers can make products that are easier to assemble. When the designers design the product and its process correctly the first time, the entire manufacturing cycle is easier, faster, and less prone to mistakes.

With a knowledge of these design techniques, you can detect overly complicated designs or processes in your factory. By suggesting solutions to these problems, you can support your coworkers and develop a team spirit. You can also keep your focus. Keeping it simple means staying in focus. The product and the process should perform their functions—no more, no less. Let's examine these techniques of simplification one at a time.

Lean Design

Most products benefit from the concept *less is more*. A product that performs the same function but takes up less space, weighs less, and has fewer component parts will usually command a higher price and be more functional. These efficient designs are referred to as *lean designs.* An excellent reference for additional information on lean manufacturing techniques is *The Machine That Changed the World* by James P. Womack, Daniel T. Jones, and Daniel Roos, Rawson Associates, Macmillan Publishing, 1990.

> **Lean Design**   Minimizes the complexity of a product by reducing the length, width, depth, weight, and number of components. Lean design reduces the number of components in a product through consolidation of functions and the use of modular parts.

A lean design has many benefits, including reduced cost for the manufacturing company. Lean designs do not come easy. They require time, attention to the requirements of the materials necessary to achieve a good design, and also attention to the manufacturing process requirements. It is much easier to design bulky, heavy products than to design light, lean products. However, the advantages are many—take Everest, for example.

Everest's customers were demanding lighter bikes, and the company decided to give its designers the challenge. The simplest approach for the Everest design team was to find lighter materials for the frame of the bike. Several materials, including titanium and carbon graphite, were evaluated until a new, higher strength aluminum alloy was discovered.

The new material allowed the bike to have a lighter frame while still achieving sufficient strength for the rugged application of mountain biking. Although this aluminum was higher in cost and required a change in the welding process in the Everest factory, the design utilized less material and resulted in a lower total cost. The use of the aluminum tubing provided more value to the product because it was leaner and more efficient. An important lesson can be learned from this example. The price of supplies and materials does not always relate directly to the total cost. Sometimes, more expensive materials will drop the total cost of a product because of the efficiency obtained with a lean design concept.

In addition to a lower cost, line workers at Everest noticed another benefit of the lean design in the manufacturing process—something quite practical. Since the assemblies were lighter (three pounds), they required less effort to move through the factory. By removing three pounds from the bike's weight, not only could the customer lift and ride the bike with less effort, the employees could move the assemblies through the factory with less effort and physical strain.

Based on the success of the new aluminum frame, Everest employees submitted 12 new suggestions on possible methods to design *lean* into other components and bicycle models. One such suggestion came from George, an Everest operator who normally kept to himself when at work. George wrote:

Employee Suggestion System

One way to improve our bike is to replace all the 10mm ring fasteners with 12mm fasteners.

If we go to the stronger 12mm fastener, we reduce the number of fasteners several places due to stronger retention force. For example, instead of using a fastener on both the front of the brake assembly and on the rear of the brake assembly, we can use one for both and thereby eliminate one whole fastener.

George

Some thought George was nuts for wanting to add weight. However, George was very observant. By upgrading the 10 mm fastener to the heavier 12 mm, one fastener could be removed completely from the assembly. Although the 12 mm fasteners were heavier, this maneuver simplified the manufacturing process through a reduced component count—the total number of different parts required to do the job. Following is the component count before and after the lean design of the ring fasteners at Everest:

Ring Fastener Analysis

Standard Design		Lean Design	
8 mm	8	8 mm	8
10 mm	12		
12 mm	**16**	**12 mm**	**24**
Total	36	Total	32

With the lean design, Everest succeeded in reducing both the part count of fasteners from 36 to 32, and the number of different types of components from three to two. By incorporating George's idea, Everest saves in several ways. It saves in the amount of inventory it carries because there is no need for 10 mm ring fasteners. Also, the plant does not need to carry 10 mm tools. More importantly, there are four fewer fasteners in the final bike, therefore, the assembly process is easier, and the finished bike is lighter.

Generally speaking, everyone benefits if the designers can achieve the objective of fewer components. The more components in a system, the more potential problems can result over time as the parts age. Screws will loosen, metals will rust, and motors will fail. With more parts, there is more probability of failure. Since most products that are assembled in factories involve more than one component or part, there is usually an opportunity to reduce the number of parts in any design. The smaller the component count, the easier it is to manufacture the product.

If you are building identical products, and your product has 150 components while your competitor's product has 100, who will be able to assemble the product more easily? Who will have to buy and store fewer parts and less material? Whose product will be lighter, easier to move, and more reliable? The answer is obvious. Fewer parts simplify the manufacturing process.

Achieving fewer parts with a lean design technique is not restricted to a single component, such as a ring fastener. By sharing product designs across several applications, the designer can also reduce the total part count in a product. In the case of Everest, a lean design could be the use of a *modular* brake assembly that is designed into all styles of bikes. This eliminates special brake assemblies for each style of bike.

With an automobile, a modular design may be a standard engine that is used for more than one model of car. In any event, by providing interchangeability of components you can also reduce the component count, and add value to the product.

Lock and Key Design

This design technique sounds like it looks. The name refers to how two different parts come together and match up. The lock and key design results in components that can be assembled only one way.

> **Lock and Key Design**   Assures that critical components can only be assembled in one direction, thereby preventing incorrect assembly.

This type of design is used primarily for applications where two parts have the possibility of being assembled in the wrong direction or orientation. When this occurs, the manufacturing problems can be serious. Everest experienced problems in its wheel assembly area because a bearing was placed 180 degrees out of alignment on at least one bike a week. Even though the employees were well-trained, it was difficult to see the difference between the top side of the bearing and the bottom side.

Occasionally, a mistake would happen. When a bearing was assembled incorrectly, the bike had to be disassembled and reworked. The problem was costly and irritating because of the wasted time, materials, and energy. Everest's first move to correct the situation was to add an inspection step after the wheel assembly to check for misplaced bearings. This inspection technique was acceptable, but added extra time and cost to the manufacturing cycle.

The same team that worked on the fastener problem got together and closely studied the bearing design. They analyzed all the different methods of assembly that could cause a mix-up between the bearing and the housing. Figure 7.1 shows two possible assemblies of the old design.

As you can see, with the old design the bearing can fit into the housing two ways—correctly and incorrectly. With a *lock and key design*, it would be impossible to put the bearing in incorrectly, because of the addition of a stop mechanism that prevents improper placement. Now look at the bearing redesigned with the help of an Everest employee involvement team as illustrated in Figure 7.2.

It is impossible to put the bearing in upside down because the flange prevents the bearing from being inserted. With a lock and key design, mistakes are greatly reduced, and the ease of assembly is improved. Although the lock and key design method is valuable, it shouldn't be considered a cure-all method for preventing improper assembly. If you

FIGURE 7.1

Standard Bearing Design

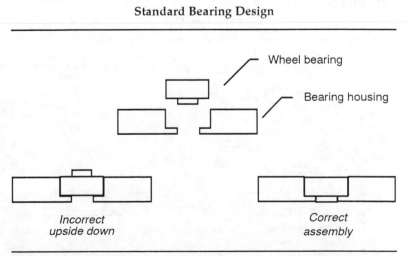

Wheel bearing

Bearing housing

Incorrect
upside down

Correct
assembly

designed every part in a product with the lock and key method, it would cause problems.

The reason is simple—many parts in a product are best left symmetrical. Take a simple washer that is used on a bolt and a nut. If the washer actually had a top and a bottom to it that were different, it would take twice as long to insert the washer over the bolt because you would first have to orient the washer to the correct side. There is no sense having washers that only go on one way. It is best to use lock and key sparingly in those designs that require a prevention device for incorrect assembly.

VALUE ANALYSIS OF THE PROCESS

Although you have learned two valuable design techniques that will improve the manufacturing process considerably, you still need to examine the manufacturing processes closely to determine if there is any waste in the system. Are the

FIGURE 7.2

Lock and Key Design

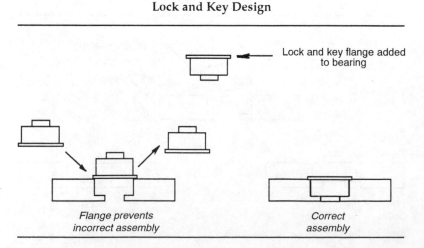

Lock and key flange added to bearing

Flange prevents incorrect assembly

Correct assembly

processes simple and straightforward? If you have a lean design, the process will be simpler since it will require fewer components and less manufacturing space for assembly. However, waste can get into the manufacturing process even with a lean design. Waste—the extra steps that do not add value to the product—must be removed from the system. A manufacturing flow chart illustrates in graphic terms which steps add *value*, and those that don't. Everest used the flow chart in Figure 7.3 to evaluate its manufacturing process.

There are very few *value-added* steps performed in this department. When the product is moving from place to place, or when the product is being inspected, value is not added. Instead, waste is added through extra dollars, time, and effort. Of the 13 operations in the department, only 4 add value! For example, attaching the gears adds value to the bicycle. A process flow diagram helps you see where the value is added in the operation—the rectangles—and where the wasted steps are—geometric shapes other than rectangles. Everest analyzed all the nonrectangle boxes and carefully worked with the product designers, the manufacturing engineers, and the line workers to eliminate nonvalue-added steps, including the unnecessary transportation, inspection, and rework steps. Figure 7.4 on page 97 is the final flow chart after a few key changes were made to the design and the process.

The results are remarkable. Instead of having 4 operations out of 13 adding value, Everest now has 3 out of 6 operations adding value. Which do you think will be more cost-effective? The advantages of a process flow diagram are obvious. The waste built into an operation is easily seen and corrected. This technique of analyzing for value is not limited to the manufacturing process; it can be used for design analysis, material analysis, and product analysis.

FIGURE 7.3

Process Flow Chart—Bearing Assembly

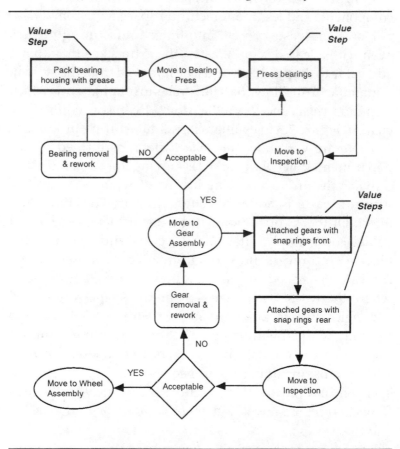

Design for Manufacturing and Assembly

Manufacturers recognize that the design of a product has a strong impact on the manufacturing process and vice versa. After production has begun, it is difficult to adjust the process. If the product has too many parts and is too complex and difficult to manufacture, the design can be a costly mistake.

FIGURE 7.4

Process Flow Chart Bearing Assembly After
Value Engineering

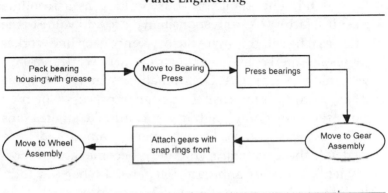

The design and manufacturing process can be developed simultaneously. In this way, more individuals in the factory can be involved in the design of the product and determine its manufacturability. By analyzing the impact on manufacturing through proper planning, the materials, tools, and basic design of the product will be greatly enhanced. The technique of developing the product design and the manufacturing process simultaneously is called *concurrent engineering*.

> **Concurrent Engineering**   The design, the manufacturing process, the materials, and the manufacturing tools are worked on at the same time to reduce manufacturing cycle time and increase the likelihood of trouble-free manufacturing. *Synonym—simultaneous engineering.*

Concurrent engineering is a common-sense approach that involves not only design engineers, but also key manufacturing personnel and sometimes customers in the initial product design. It is a form of engineering that is not isolated; it encourages communication. A part is not simply designed in a vacuum, but rather relies on team input. Why is this important for your position as a frontline worker in a factory? Your responsibility is to add value to the product as it travels through a factory. Since each line worker becomes an expert in his or her department or work area, you have a unique perspective. Occasionally, you may have ideas that can cut costs or improve the design or process. You may not be asked your opinion, but you can make suggestions anyway. After you have been properly trained and have gained valuable experience, your ideas become more important. Yields can easily approach 100 percent when key individuals in a factory get together and design a product that can be built easily and correctly. This process of combining manufacturing expertise with design engineering is called *Design for Manufacturing and Assembly (DFMA)*:

Design for Manufacturing and Assembly (DFMA)
Requires that designs be compatible with standard manufacturing processes, thereby achieving high yields and quality product during the manufacturing cycle.

By involving supervisors and line workers in the design of new product, the likelihood of a good design and a successful manufacturing process is greatly increased. The principle of DFMA translates to planning ahead and anticipating those designs that will flow through the factory with higher yields. These include modular designs, designs with fewer components, and designs with foolproof assembly mechanisms that eliminate inspection procedures.

DFMA involves a good understanding of the relationship between the design, the material, and the process. A good design fits the factory like a glove. When the design is right and is compatible with the manufacturing process, we refer to that design as *robust*.

> **Robust Design**   Design that is easily manufactured because of its simplicity and its excellent compatibility with the manufacturing process and the materials used.

This is the ultimate design—a product that slips easily into a manufacturing factory. Everything works correctly the first time. How do you know when you have a robust design? Most operators will know right away. The tools and machines are available to do the job easily and quickly. There will be fewer parts, and the parts will fit the first time without rework. Fasteners will be well-placed, with plenty of room for positioning on the work surface. A robust design will produce an excellent final product without wasting materials and effort. Good designs are simple and obvious when you understand the concepts of lean design, value analysis, and lock and key.

TIPS

(1) Understand the relationship between design and process. A design is only *good* if it can be manufactured with reasonable yields. Involve yourself in design review meetings whenever possible, and put your manufacturing know-how into the new designs your company is building. Remember that many ideas are rejected for good reasons. If you have one good idea in fifty, be proud of it, you are doing better than most.

(2) Lean design has many hidden benefits from a control standpoint. With fewer parts to assemble, the process is easier to control. When you see an opportunity to reduce the number of parts, speak up and suggest your idea.

(3) Suggest to your team that a value analysis study be performed on your manufacturing process and your design at least every six months to prevent waste from creeping in. Are you moving your product excessively; inspecting it excessively? Is every component and piece of material in the final product necessary for its integrity? Will excessive parts cause premature failure?

(4) The *keep it simple* concept not only applies to the design and the process, it also applies to your paperwork and your documents. When reporting your activities, place the most important information on the first page, and avoid excessive explanations. Get to the point, and write clearly and concisely.

CHAPTER REVIEW

(1) Look up the term *design for disassembly* in the Glossary. Would you expect to see this type of design skill used in a concurrent engineering setting? Is it environmentally sound?

(2) Everest considered consolidating some rooms in its factory—taking 15 separate manufacturing rooms and removing some walls to form 10 rooms. Would this improve operations? Use a value analysis flow chart to explain your answer. Does the movement from room to room add value?

(3) Draw a process flow chart for your work area. How many added value steps are there? How many wasted steps are there? How can this be improved?

(4) Is *move to inspection* a value-added step? How about *bearing removal and rework*?

Chapter Eight
Daily Improvement

Rule Eight

Continuously improve the manufacturing process and the product.

S uperman® (a registered trademark of DC Comics, Inc., a Warner Bros., Inc. company) is an instantly identifiable American folk hero. You can probably visualize him dodging bullets, picking up locomotives, and leaping tall buildings in a single bound. Unfortunately, in real life it isn't that easy to move trains and jump over buildings. However, there *are* ways to accomplish extraordinary feats if the task is divided into smaller more achievable goals. These tasks can be performed with a simple concept called *daily improvement*.

Interestingly, teams of people who make small movements in the same direction can accomplish substantial goals that may seem out of range for any one individual. Take sports, for example. A basketball team beginning its first year in the junior varsity has a long way to go to develop the personal and team skills necessary to perform at a varsity level. However, through daily practice, a competent basketball team can conceivably begin to win games within a year. This will not happen by spending two days at the gymnasium in intensive training. It happens over time with small amounts of continuous improvement.

Accomplishing goals inside a factory is the same as improving the skills of a sports team. In a factory, you have approximately 250 days during the year where you work with a group of individuals manufacturing a product. If you improve a little each day, the result at the end of the year will be substantial. If each individual within the factory strives for daily improvement, the end result is truly awesome. A manufacturing facility where each of the 100 employees makes one small improvement each day would make a big difference. One-hundred small improvements times 250 days equals **25,000 improvements a year!!!!!**

A small improvement when added to other improvements can produce incredible results. If you are observant, and have a clear picture of what you want to accomplish, you can make small improvements toward your ultimate goal—high quality product. Occasionally, a significant improvement occurs through a new innovation or an idea, but you should prepare yourselves for small, continuous movements upward. Inch by inch, it's a cinch! This journey of continuous improvement is described in the Japanese language as *Kaizen*.

> **Kaizen**   Japanese term for daily continuous improvement. Statistical methods based on the teachings of Deming are utilized in this system to reduce variability in the process and the product. *Synonyms: continuous improvement process; incremental improvement.*

Americans tend to view the world differently than the
Japanese. We enjoy hitting the home run in the ninth inning
and scoring big on a new product. We tend to like the big
leaps—the type of innovation that revolutionizes an industry. The Japanese became students of our manufacturing
processes after World War II and have done an excellent job
of learning. It is now time for the teacher to learn something
from the student. The idea of small improvements every day
that add up to substantial improvements is a powerful concept that we, as United States manufacturers, need to adopt.

Continuous improvement requires attention to detail and
perseverance. As a frontline worker or supervisor, each day
should be approached with the attitude that even a small
improvement in the factory is important and will add up to
big gains. If you use any of the techniques described in earlier
chapters, you can make daily improvements. Take spit and
polish for example. If you improved your equipment details,
your work area, and your product, just a small amount every
day, you would not have significant cleanups or challenges
facing you at the end of the week or the end of the month. You
would also increase the likelihood of quality product.

HOLDING THE GAINS

Before introducing additional methods for continuous improvement, you must learn to preserve the gains that you
have already made. The foundation of Kaizen is building
upon successes. What is the opposite of continuous improvement? It could be called *continuous falling apart.*

This falling apart syndrome is experienced by many organizations that fail to fix chronic problems. They eventually
degenerate and go out of business. The plant gets dirtier with
time; the product lacks detail over time; procedures are lost or

not followed. Sometimes, these chronic problems are blamed on a famous guy who's known for showing up wherever there's bad news.

> **Murphy's Law** A common term that refers to the adage: *What can go wrong, will*, Murphy's Law describes the degenerative process that often accompanies manufacturing processes that lack adequate controls and planning.

Murphy has never been seen, but he is famous nonetheless. If there is a possible way of failing, Murphy apparently seeks this method out and tries it the first chance he gets. Murphy tends to disrupt activities—everything from machines breaking down at the worst possible moment, to organizations literally falling apart. Everyone tends to joke about Murphy's Law. The concept behind Murphy, however, is real: Murphy attacks *organized* objects and groups of individuals due to a particular scientific law—organizations tend to degenerate with time. Scary stuff—but there is an answer to this problem. Murphy is a natural effect of nature, and he can be controlled if you become more organized and learn from past mistakes and failures. The key to controlling Murphy is *process control.*

> **Process Control** Uses feedback loops, *statistical process control (SPC)*, and *standard operating procedures (SOPs)* for the maintenance and adjustment of all manufacturing processes.

You may recognize some of this definition, since we defined statistical process control earlier in the book. This definition of process control is a little broader in scope and attempts to describe the control needed for all processes within a factory. Through process control you can achieve a *bullet-proof* organization that consistently manufactures quality product.

Process control is essential to any manufacturing activity. It is the system or systems that monitor your direction, your product, your general activities—and send feedback signals to you when you start to go off track. If Murphy's Law is starting to act on your manufacturing system, a good process control system will alert you that this is going on so you can correct it quickly.

An example of Murphy's Law is equipment wear and tear. If equipment is not maintained, it will break down quickly. This *wear* is Murphy's Law. A machine is an organized system of parts that will degrade with time and cause problems. There is no such thing as perpetual motion in manufacturing. Equipment, tools, and facilities require periodic care to stay organized and operating properly. A good process control system does exactly that. With process control, a signal goes out, either through an audit, a procedure, or an employee's observation that a particular machine, process, or employee needs improvement. This signal and the response to it become the process control that keeps everything in the factory on track and performing its function.

Since things are constantly changing, process control must be a dynamic process. Unfortunately, people leave companies for other jobs, suppliers sometimes go out of business, and it is human to make mistakes or forget procedures. For these reasons, a good process control system permanently stores knowledge that can be protected and built upon. This permanent storage of knowledge can reside in computer storage programs, training programs, documents, and employee notebooks. It is important to hold on to whatever you have already learned.

Some large companies tend to periodically relive problems due to a constant loss of the information and skills that they had acquired and built their success upon. Many companies, however, are becoming smarter about storing the knowledge of the business and periodically publishing it so the factory can be checked and audited to verify that it is on track.

> **Memory Retrieval System (MRS)**   This is a computer listing of critical factory parameters, including standard operating procedures and specifications that have been stored and are periodically displayed for audit by individuals within the factory.

Regardless of how the information is retained and updated, it is important to *hold the gains*.

> **Hold the Gains**   Hold or retain the business knowledge, technology, training, and other intellectual information developed and acquired during the growth of a manufacturing firm.

The above definition relies on basic common sense: you should hold on to what you know, and use it wisely. Knowledge of your basic process is the true foundation of continuous improvement. Holding the gains is a skill that constantly challenges us. Pride of workmanship, spit and polish, periodic maintenance, doing it right the first time, and other techniques described in this book are vital to holding on to what you have already learned about your factory. These *mental push-ups* will stimulate an awareness of where you have been and where you are going. Using these techniques in your day to day activities can lead to exciting, challenging workdays.

THE BASIC FEEDBACK LOOP

The *basic feedback loop* is the crown jewel of manufacturing controls and systems. It should be understood, studied, experimented with, and utilized in every aspect of your daily activities. Know it, and know it well. Keep it as your secret weapon—it is a key to continuous improvement.

The first feedback loop described in this book was the Just-In-Time (JIT) concept. The JIT feedback loop works by signals and response to those signals. A manufacturing operation, for example, sends a signal to a previous operation when it is running low on inventory and needs more product sent in. This signal triggers the previous operation to manufacture more product and send it forward. Feedback loops involve a continuous set of signals followed by actions to the signals.

> **Feedback Loop** Manufacturing information transferred backwards to the previous work station that signals the need for action or improvements in the process.

A feedback loop provides information about a process in a reasonable length of time. This backward information flow provides valuable insight into how an operation is running, and whether or not it needs adjustment.

Feedback loops are all around you in a variety of forms in your houses, your communication patterns, and even in the operation of your body. For example, when you drive down the highway, you turn the wheel to adjust your position within the traffic lane. You would run off the side of the highway if you did not receive feedback from your eyes as to your position in the lane. By seeing your position while correcting your course, your mind interprets the correct time to stop turning the wheel. Your sight is part of the feedback loop that signals your brain to stop the adjustment of the steering wheel. A feedback loop consists of a signal followed by an action. Your action is to turn the wheel after your eyes have generated the signal to correct your original course.

A feedback loop in manufacturing is not much different, and in the next two chapters, some more sophisticated forms of the feedback loop will be introduced. The more feedback loops in a factory, the more control there is in the manufacturing process. If the feedback loops have a good signaling method that is sensitive and accurate, the signal will carry significant information about the process that will allow small, gentle incremental corrections to the original process. The signal is generated by a *measurement* of the process. Therefore, the more accurate the measurement, the more accurate the signal transferred backwards to the process.

In the driving example, the human eye is an extremely sensitive instrument that provides our brain with small accurate signals concerning the adjustments that will bring us back onto the freeway lane. These small adjustments are exactly what we were looking for in this chapter—incremental improvements—Kaizen.

Your goal is to achieve a manufacturing process that holds its gains, while improving in small amounts over time. In order to achieve this, you need to know exactly where the process is at any point in time. Feedback loops will be your guide, and they will be used at many different levels.

Feedback loops provide valuable information that can be used to improve your product, process, or performance. You are an expert—the best judge of the quality of the process and product that move through your department. You make judgments about the process and product through your sense of sight, hearing, touch, and smell. When something doesn't look or sound right, you question it. You also make judgments about the process or product by collecting and analyzing data. You take corrective actions to improve the quality of the process or product based on your judgments. By continually measuring your process, and then adjusting the process in a manner that improves the product over time, you can continually improve.

Figure 8.1 below is a diagram of a feedback loop that provides data that are collected and analyzed prior to making judgments and taking corrective actions.

The diagram shows one of several loops inside the Wheel Assembly department. This is the Tire Gage Inspection loop. An operator, Mike, is responsible for measuring the tire pressure after wheel assembly and recording it as a variable measurement on a process sheet. (Do you remember the difference between variable and attribute measurements?)

In this specific case, the initial tire pressure was too high, which was Mike's signal to take action and adjust the tire

FIGURE 8.1

Local Feedback Loop

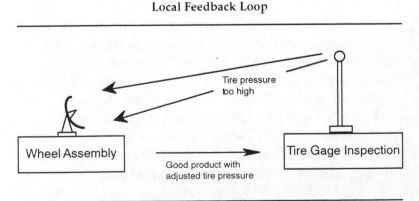

pressure to the correct value. Mike knew an adjustment was needed when his measurement indicated that the tire pressure was outside of the specifications contained in the standard operating procedure. Mike held the gains when he measured his work and corrected the problem.

Good feedback loops exist throughout the factory and extend to the customer. When a customer finds a defect in your product, it is a signal for corrective action. If Pete's Bike Shop (one of the end customers) has a problem with tire pressure, the feedback loops should transmit that information backwards from the bike shop to the Shipping department, to the Final Assembly department, and so on, until the problem is corrected.

The feedback loop system, which is the heart of process control, is not just a system for holding the gains. It is also the technical side of the NOAC concept we learned previously. Learn to recognize the signals that will help to ensure that you pass on good product. Feedback and corrective action are the basis of a good supplier-customer relationship. You need your customer to tell you how you are doing; your customer provides valuable signals to help correct your course and adjust your process. Feedback systems are the basis of process control and the basis of holding the gains. More importantly, if you can fine tune this system of adjustment and course correction, you can go beyond control, and actually improve your process over time.

PLAN, DO, STUDY, ACTION

Dr. Deming, whom we learned about in Chapter 1, has been a leader in manufacturing management concepts, including process control and improvement. His feedback loop concept is termed *Plan, Do, Study, Action (PDSA)*.

Plan, Do, Study, Action (PDSA)   The Deming methodology of planning an operation correctly, carrying out the operation, studying the operation for its performance, and correcting the operation with an action if required, is a mature feedback loop for manufacturing control and continuous improvement.

The plan in Dr. Deming's cycle is the total blueprint of the procedure, including the design, the material, and the process. Do is the action of carrying out the manufacturing process. Study is the measurement and analysis of the manufacturing process, which usually includes the measurements of attribute and variables of both the process and the product. And finally, action is the corrective action required to bring you back to your plan. This PDSA cycle is a mature version of the feedback loop.

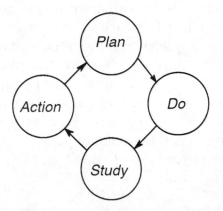

The plan provides direction, and the other steps are an active process of performing, collecting, analyzing, judging, and correcting. It is a dynamic process that relies heavily on the use of your intelligence. If you follow this formula, you will be working smart and continuously improving.

A SYSTEMS APPROACH TO CONTINUOUS IMPROVEMENT

Since the science of manufacturing is relatively new, there aren't many integrated systems that tie together all the concepts of process control and continuous improvement. Usually, manufacturing systems emphasize either Total Quality Management, or a single concept such as Just-In-Time. Eventually, integrated programs will be developed that emphasize all the skills necessary to be balanced manufacturers. General Motors, Ford Motor Company, Chrysler, and Motorola are a few companies that already have excellent programs on the leading edge of integrated systems. These manufacturers understand the importance of process control and continuous improvement, and they require a systems approach inside a supplier's factory. The systems cannot be superficial, but require substantial planning and employee involvement. Each process must be well-defined, and a process *plan*, called a *control plan*, that defines all of the sources of variability inside the supplier's plant, must be in place describing every feedback loop in the factory.

> **Control Plan** Defines all of the primary product and process feedback loops in a factory and contains a *reaction plan* that describes in detail the action required on all feedback signals.

In addition to the control plan, other documents called *Failure Mode and Effects/Analysis (FEMA)* and *Design Failure Mode Analysis (DFMA)* are required in mature manufacturing facilities. These documents describe what to look for when something goes wrong, or what the consequences are of something going wrong inside the plant.

Failure Mode and Effects/Analysis (FEMA) Lists potential causes of failures in a manufacturing process and the possible consequences of each failure on the end product.

Design Failure Mode Analysis (DFMA) Studies all possible failures of a product and the possible consequences of each failure. Evaluates how much damage *Murphy's Law* can create in a product.

A FEMA and/or a DFMA are more sophisticated versions of a fishbone diagram, because they not only list the potential causes of a problem, they also define what the consequences are if the problems are not fixed quickly. These systems are extensions of feedback loops and holding the gains. It is not easy to be a supplier to GM, Ford, Chrysler, or Motorola. These manufacturers have strict requirements and demand that high quality materials go into their product. Although your factory may not contain a comprehensive system for process control and continuous improvement, you can almost bet that one will be adopted in the near future. Such systems are a must if you want to remain competitive. Remember, inch by inch, it's a cinch!

TIPS

(1) Learn to recognize and use the feedback loop in as many applications as possible. It is your way of steering the correct course of action.

(2) Data obtained from instruments and counts are often accepted at face value. Always challenge your data because you make important decisions based

on it. A display on a computer report, an instrument gage, or a digital readout, don't guarantee accuracy.

(3) Mistakes are often repeated because factory *memory* was not checked, updated, or applied. Use and update the memory for continuous improvement.

(4) Feedback loops are more effective when the time interval between this measurement and the action is short and frequent.

(5) It is helpful to record in a *process log book*, (which you may have already started), the successes or failures that have occurred during your quest for continuous improvement. Just as you normally keep a checkbook up-to-date to determine the balance of your bank accounts, a process log detailing events will chart your progress and refresh your memory.

(6) Add some reading on process control to your agenda. It is a subject with depth and numerous applications.

CHAPTER REVIEW

(1) Is an employee newsletter a feedback loop? Explain your answer.

(2) Would a semiannual employee newsletter be better or worse for employee communication than a monthly newsletter? Why or why not?

(3) Describe a Deming PDSA cycle already in existence in your factory by listing the plan, the do, the study, and the action. Be specific on each item.

(4) Name two personal activities where daily practice improves your skills.

(5) Locate an article in the *Business* section of today's
    paper that either describes a company's continuous
    improvement program, or cites an instance of
    Murphy's Law striking out! How long has the com-
    pany been in business? How long do you think the
    company will remain in business?

Chapter Nine
Reducing Variability

Rule Nine

Continuously reduce the variability in the process and the product.

It is virtually impossible to have two totally identical things. Identical twins may at first glance appear identical, but after careful examination, differences become more obvious. Nature resists the production of exact sameness. Similarly, manufacturing strives for sameness, but will always be plagued by variation. Upper and lower specification limits are routinely established to cover a range of variation since exact sameness cannot be produced. Products that fall within these limits are similar enough in looks and function to be acceptable to customers. Generally only sophisticated instruments can detect this slight variation. Acceptable variation within specification limits is referred to as *natural variation.*

Natural Variation   Tendency of similar products, materials, and processes to be slightly different within a
certain range of values.

The goal of manufacturing is to produce products with as
little variation as possible. When the process departs from the
specification limits, the variation is too great. You are no
longer producing similar products. This is referred to as
uncontrolled variation and will probably result in defective
product, increased scrap, and poor yields. An uncontrolled
variation may be the result of an accident, but more frequently it is the result of someone not following procedures.
Variation can usually be traced to human error, faulty machines or equipment, or defective material.

Uncontrolled Variation   Variability due to unplanned
events such as human error, faulty machines, or defective material.

You can find and correct the cause of uncontrolled variation, but first you must become aware that it is present. An
uncontrolled variation in a factory is the most common
reason for defective product. Your goal is to reduce uncontrolled variation in the manufacturing process and product
through planning, process control, and tracking uncontrolled variation back to its source—the *assignable cause*.

AN ASSIGNABLE CAUSE

Variation is caused by something or someone. When you
identify the cause of uncontrolled variation, you have identified the assignable or special cause. For example, Americans
live an average of 77 years. Some people live more than 77

years, and some people live less than 77 years. Nature is the
cause of this variation, so it is referred to as natural variation.
People who smoke have a shorter life span. The lower average
age for smokers is not due simply to natural variation; it is the
result of smoking. The shorter life span can be traced to an
assignable cause—smoking. Variation that can be traced to an
assignable cause is usually related to human error in one form
or another.

> **Assignable Cause**   The cause of uncontrolled varia-
> tion in a manufacturing process. For example, a power
> failure may cause scrap and defective product; it is the
> assignable cause for the scrap and defective product.
> *Synonym: special cause.*

If you have a flat tire on the highway, and find a nail in the
tire, the assignable cause is straightforward—the nail. The
nail is not part of the normal road conditions and is a special
circumstance that caused the flat tire. Determining the assignable cause is not always that simple since there may be more
than one assignable cause, or the cause of the problem may be
several steps removed from the symptoms of the problem.
For example, a tire may become flat from a defect that was
manufactured into the tire and *telegraphed* through the manufacturing process to you. These assignable causes can be
found earlier in the manufacturing process, before they cause
too much damage, through the technique of *charting*—a
method of clearly illustrating variability.

Charting Variability

Charting is one of the best methods of detecting natural
and uncontrolled variation. The goal of process control is to
stay between the upper and lower specification limits in order

to maintain natural variation. This is like driving on a free-way—if you swerve all over the road, you will get into an accident. You want to stay as close to the center of the lane as possible. If you learn to detect the signals that warn you when you wander off course, you can adjust the process, and get back on track. Figure 9.1 illustrates tire pressure variation during a typical day at Everest.

The straight path has been disrupted by variability. Corrective actions reduce variability and keep us within limits. The excessive variability noted on the chart is our call to investigation and action. The chart is a source of feedback. If the operator responds to the action signals and corrects his or her course, the product will again be within specification. By detecting the *action signals* that coincide with excessive movements away from the center line, the process can be corrected and smoothed. Over an extended period of time, a continuous response to these action signals will result in a straighter path with less variability.

Charts can be used to illustrate data from a variety of sources. For example, the pressure from both the air pump

FIGURE 9.1

Charting a Course-Tire Pressure

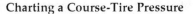

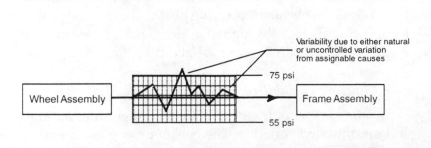

and the tires can be measured, charted, monitored, and adjusted. The goal of this chapter is reducing variability. This is not restricted to any one department or any one measurement. Rather, it is an attempt to reduce variability in all aspects of the product and the process. This goal is achieved by responding to the action signals and steering the process back on course.

PROCESS CONTROL ACTIONS (PCAs)

It is your responsibility to act when the data in the feedback loop reveal movement away from the center line of the process. As the frontline worker or supervisor, you need to respond to the action signals, and record your attempts to move the process back into line. All corrective actions should be documented by either writing a report, writing on the chart, or using a process control action form.

> **Process Control Action (PCA)** Documented adjustment in the manufacturing process that reduces variability and is prompted by an action signal.

PCA is a way of documenting an action on the process. It is a useful tool that is also a reference for the causes of variability in the process. If you adjust the process, you must record it. The PCA contains a description of the problem, a description of the cause, and the corrective action taken. The PCA provides continuity for other coworkers and supervisors, and it becomes a valuable information base that allows further improvements to be performed on the process without losing ground. There are numerous forms of PCAs. They can be handwritten responses on charts in the manufacturing area, or separate documents that circulate inside a factory.

They can also be entered into a computer. The form itself doesn't matter as long as the action took place, and the action was recorded. Figure 9.2 shows how a PCA system might work to reduce variability in the Everest factory.

You can now observe that the variability in the process has been reduced substantially—notice that the black line doesn't wander as erratically as before.

The PCA system is a powerful concept that was developed at Sigma Circuits in northern California as part of their *CITE* continuous improvement program. Improved quality levels and improved customer satisfaction were achieved in a short period of time at the Sigma operation. The PCA system was not an outside system developed by *experts*, but rather an innovative idea that evolved from the active participation of all the Sigma employees—from operators to managers—in a real team approach. The PCA system is a simple, common-sense method of improving the process and provides documentation for further analysis at a later date. The results generated by the Sigma factory employees are no different than the results you can generate in your own factory.

Control Charting

Figure 9.2 is an example of a control chart, which should be studied more closely to understand its significance in reducing variability. Control charts are used extensively throughout factories to reduce the variability of a product and a process. In a typical factory, there can be hundreds of measurements that would be suitable for control charting. At Everest, for example, the Wheel Assembly department actually has 30 measurements in the assembly process—tire pressure is just one. All are charted and must be within a certain specification range. Everest uses control charts as a graphic representation of variability.

FIGURE 9.2

Process Control Actions (PCAs)

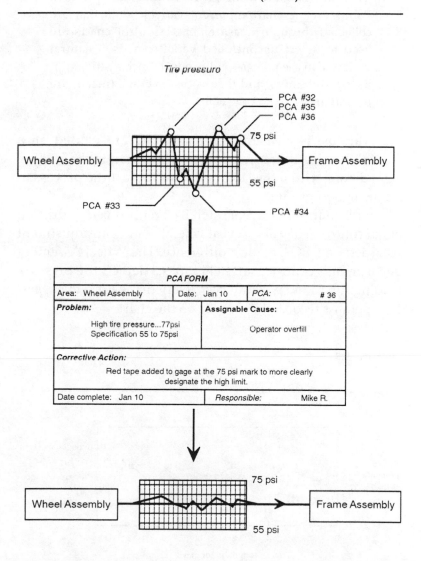

Control Chart   Graphic representation of the variability in the manufacturing process and/or the product.
Points on the chart represent variable data that are
collected through measurements. Control charts are
used to detect uncontrolled variation versus natural
variation in the process and product. Specification limits, control limits, and the process average (mean) are
usually indicated on the chart.

A variable is measured, and the data are plotted on a
control chart over time. Reading a control chart is like reading
a good book. Often the story tells us to adjust the process to
achieve good product.

Trends in the control chart tell us if our process is drifting
out of control. They also reveal whether or not an adjustment
recorded on a PCA made a difference. The Wheel Assembly
department's control chart in Figure 9.3 depicts tire pressure
readings on Monday, January 8. Every tire was checked, and
the operator recorded the value on the chart.

FIGURE 9.3

Control Chart / Monday

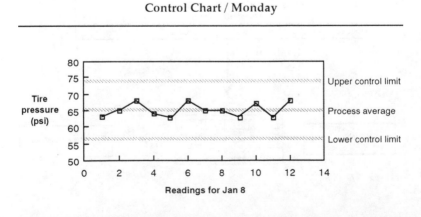

Notice that all of the values fall well within the control limits. *Control limits*—a new term that is slightly different from specification limits, which we discussed earlier—tell you what a process has been running like for the last several weeks. Control limits can be calculated many different ways, but normally represent a range on the chart where there is a 99.7 percent probability that the next measurement will fall. Since there are a number of excellent reference materials on statistical calculations, such calculations will not be addressed in this book. However, we will explore both the basics of reading charts and the methods of determining when to adjust a process and when to leave it alone.

For example, the trend line in Figure 9.3 is relatively stable and close to the center. This type of pattern is a visual clue that the process is in control and doesn't need any adjustments or PCAs. The process is *idling* in position and exhibiting natural variation. Natural variation is usually apparent when the movement around the center is relatively smooth, without any significant movements up or down. This natural variation is expected, and your product should be good. So why waste time recording the pressure readings if you are only seeing nature at work? It sounds like a lot of trouble. Maybe, but look at Tuesday's chart in Figure 9.4.

If you learn how to study the control chart, you will see something very interesting. Tuesday's chart indicates a potential problem. How can that be? All of the readings are within the control limits, and the trend line is still relatively smooth. However, if you look closely, there is a trend upward starting around the sixth reading of the day. The process is trending out of control. Could there be an uncontrolled variation somewhere in the factory? The trend is toward high pressures, and it looks like tomorrow there could be some real trouble. Look at Wednesday's readings in Figure 9.5.

FIGURE 9.4

Control Chart / Tuesday

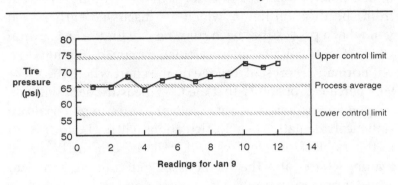

FIGURE 9.5

Control Chart / Wednesday

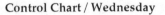
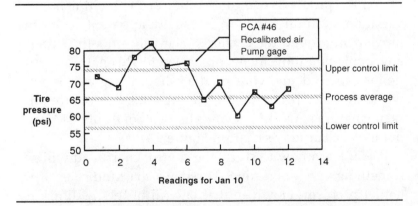

The tire pressure exceeded the control limits set up from previous runs. The process appears to be careening out of control—something like a drunk driver on a highway who swerves into other lanes. The chart signaled the problem on the sixth reading Tuesday. When you notice a trend, it is your signal to investigate and take appropriate action.

It is your responsibility to take action and bring the process under control when you detect a signal for action, such as the high value on the sixth reading on Wednesday. (See Figure 9.5.) Knowing what an action signal is in its various forms takes training and attention to detail. At Everest, the operator recognized a troubled situation and decided to fill out a PCA on the sixth reading of the day.

Sure enough, the Everest operator made a difference. By watching the chart and responding to the action signal, he was able to bring the process back into control. If the operator had been more experienced at reading the chart, he could have brought the process back into control eight hours sooner. However, he will have to find the bikes that were built on the morning of January 10 and release some of the air from the tires. (Those bikes have tires that have pressures above the specified limit of 75 psi.) That's going to cost some money and take some time!

READING THE TRENDS FOR ACTION

There are certain trends that an operator should look for that signal uncontrolled variation and the need for action. It is important to find the assignable cause and take corrective action. When reading charts, watch for the following:

- Seven consecutive points above or below the process average
- Seven consecutive increasing or decreasing points
- Any point beyond either control limit
- Any point beyond either specification limit
- Ten out of eleven points on one side of the process average

If any of these trends are spotted in control charts, it is imperative that you react and correct the situation by adjusting the process toward the center line. It takes time to learn how to read trends in graphs. It is best, therefore, to count your points each time you place a new point on the graph. In this way, it is unlikely that you will miss an opportunity to investigate and correct the process.

Control charts help you keep the process and the product within specification. Up to this point we have talked about variable data. We will now look at the control chart for attribute data which is a P chart. You should become comfortable using charts for both attribute and variable data because both types of data are important for reducing variability.

P Charts—Easy Signals for Action

An attribute is characteristic of a product or a process that is either there or it isn't. When running *live* in a factory, there is an abundance of attribute data to look for, such as product defects and product yields. Every time you have an attribute problem, it is a signal or a call for action. As operators, you should pay careful attention to any defects that occur in your department or in the material you receive from the previous department. These defects in the product can be sorted out and removed. However, if the cause of the problem is not detected, it is very likely that the problems will continue. Remember telegraphing? The best tool for plotting the attribute signals is the P chart.

> **P Chart**   Control chart for attribute data (go/no go).
> Plots the percentage of product units found to be defective over time. Space is available on the chart for control limits, as well as the inspected and rejected quantities.

The P chart basically allows you to monitor the defects as a percentage of the total number of parts that you have run for the day. On the bottom of the chart, tally the kind of defects that have occurred in the process. These signal you to action. The signals for action are readily apparent with a P chart. P charts are different from control charts because the left hand axis is representative of the percentage defective. A P chart also lists the types of defects it is tracking, such as wheel imbalance. On lot #1 in Figure 9.6 below, the P chart is showing a 2 percent defect rate. Of course, the perfect P chart would have no line on it because the process would have no defective parts. P charts are necessary tools for charting your progress toward zero defective parts and zero defective lots.

In Figure 9.7 on page 130, you can also see a control limit. Its location is based on several weeks of data. It is expected that 99.7 percent of the lots inspected will fall at or below that line, or in other words, at a defect rate lower than 2 percent.

FIGURE 9.6

P Chart for Wheel Assembly Department/Jan 10

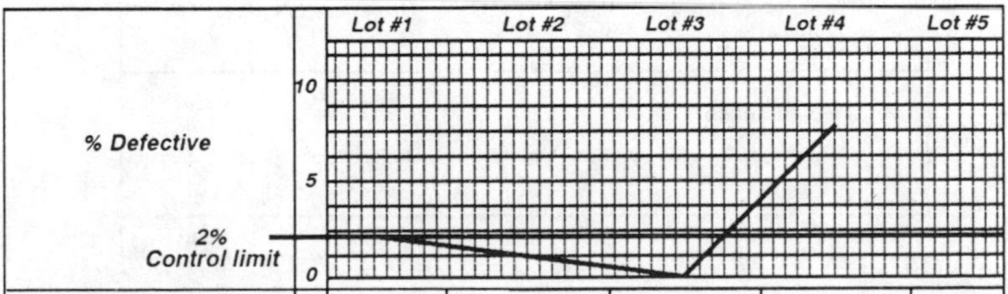

	Lot #1	Lot #2	Lot #3	Lot #4	Lot #5
Rear wheel imbalance	1	1		5	
Front wheel imbalance				1	
Incorrect bearing seating					
Squeaky wheels					
Gear misalignment	1			1	
Total units defective	2	1	0	7	
Total units manufactured	100	100	120	100	
Defect percentage	2	1	0	7	

% Defective
2% Control limit

FIGURE 9.7

The Effect of PCA #40

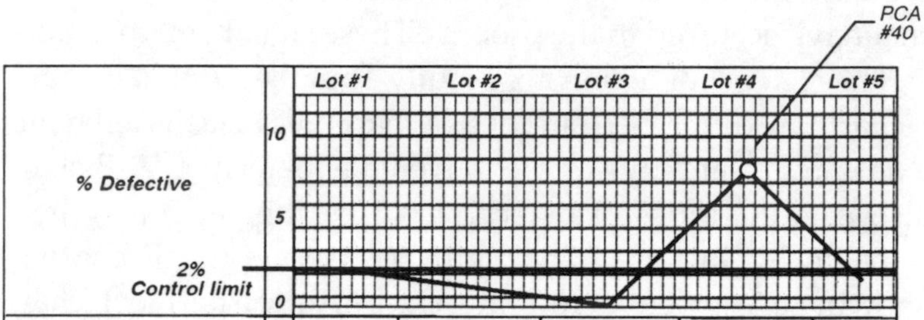

		Lot #1	Lot #2	Lot #3	Lot #4	Lot #5
% Defective	10					PCA #40
	5					
2% Control limit	0					
Rear wheel imbalance		1	1		5	
Front wheel imbalance					1	
Incorrect bearing seating						
Squeaky wheels						1
Gear misalignment		1			1	
Total units defective		2	1	0	7	1
Total units manufactured		100	100	120	100	80
Defect percentage		2	1	0	7	1

PCA Form		
Area: Wheel assembly	Date: Jan 10	PCA: #40
Problem: Poor yields due to rear wheel imbalance		**Assignable cause:** Defective rim press
Corrective action: Faulty press repaired by the Maintenance Department		
Date Complete: Jan 10	Responsible: Jean F.	

The P chart is telling you to fix a wheel imbalance problem on the fourth lot that has a 7 percent defective rate—way over the 2 percent limit. The fact that it went above the control limit is a signal for action. This time, the operator decided to do a PCA immediately and correct the wheel imbalance.

By performing PCA #40, which took action to fix a defective machine in the wheel assembly area, the Everest operator fixed the problem. This is the power of feedback loops. Lot #4 had a defect rate of 7 percent. After fixing the problem, the defect rate moved below 2 percent in Lot #5. The signal results in investigation and action. The same criteria apply for both P charts and control charts. Watch the trends and the points that are outside control limits; they call for action.

KNOWING YOUR PROCESS

Now you know how to reduce the variability in the process by reacting to trends. Can you at this point identify natural versus uncontrolled variation? In the previous examples, high tire pressure and a rear wheel imbalance were due to assignable causes. That is, something went wrong with the process and disrupted the trend lines from the normal path that was relatively smooth and centered on the charts.

Prior to the problem, the charted lines were showing a low, uniform variability. Plotted points were close to the process midpoint with approximately equal points above and below the center line. Mother Nature has obviously been at work infusing natural variation into the process. After the problem occurred, there was an unnatural spike that was not consistent with the previous trend. Remember, nature always provides some variability—natural variation—in similar items. However, a steady drift upward or downward, or a dramatic shift in the amount of variation, indicates uncontrolled variation that can be reduced. Figure 9.8 combines January 9 and January 10 tire pressure control charts to illustrate the difference between natural and uncontrolled variation. While the assignable cause was affecting the process, the trend line became more erratic and moved dramatically away from the

FIGURE 9.8

Control Chart

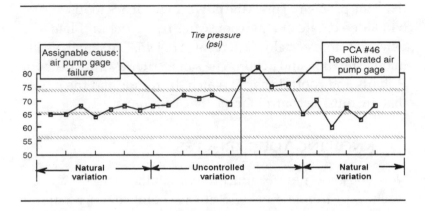

center line. When the assignable cause was corrected with PCA #46, the process returned to normal, and only the natural variation remained. As a frontline worker or supervisor, you are the best judge of how your process is performing, and it is your responsibility to control and reduce its variability.

If you can spot an uncontrolled variation, you can use other tools such as fishbone diagrams to find the assignable cause. Continue your quest to reduce the variability in your process and product through the use of feedback loops and control charting. Your next customers—your fellow coworkers and your final customer—will appreciate your efforts.

TIPS

(1) Charting your progress is a powerful tool for process control and reducing variability. It is one of the most effective feedback loops.

(2) Uncontrolled variations occasionally happen from mistakes you or your colleagues make. Don't dwell

on mistakes; use them as opportunities to make a stronger process. Use the team approach to rid the factory of assignable causes that result in problems.

(3) Don't limit yourself. Learn one type of chart well, and then explore other types of charts as you become proficient. By combining both control charts and P charts, you can reduce variability and scrap.

(4) Look for trends and unusual movements in your charts. Movement upwards or downwards for more than three to four points—or a sudden movement up or down—is a tipoff of a potential problem.

CHAPTER REVIEW

(1) Two loaves of bread are baked at the same time. Both loaves are mixed and baked exactly the same, yet one loaf is 4" high after baking, and the other is 4.25" high. Do you think the difference is natural or uncontrolled variation? Why?

(2) Another two loaves of bread are baked. This time one loaf comes out 4.5" high, and the other loaf comes out 2" high. It was later discovered that the 2" loaf was mixed without yeast. Is this difference natural or uncontrolled variation? What would be the assignable cause in this example?

(3) Which two circles in Deming's PDSA graph (Chapter 8) refer to the concept of improving the process through corrective actions such as PCAs? Where is the measurement of variation performed?

(4) When measuring the temperature of an oven, would you use a P chart or a control chart? Is temperature a variable or an attribute?

(5) A P chart is used to chart the number of paint defects in a particular Everest mountain bike. Is the chart using attribute or variable data?

(6) If a control chart suddenly shoots well above the upper control limit, is the problem due to an assignable cause or natural variation? Explain.

(7) A control chart indicates variable data in the following order: six points below the target value, one point above the target value, five points below the target value. Do the data indicate a trend? Explain.

Chapter Ten
Six Sigma

Rule Ten
Center the manufacturing process well within specification limits.

A nd now for the moment you've all been waiting for—the statistics chapter. Don't worry, the concepts in statistics are not any more difficult than some of the concepts we have already covered, such as natural variation and JIT. Since there are several good books available on the mathematics of statistics, and you can even buy calculators that do the math for you, this chapter will concentrate on the principles behind the statistics instead of the mathematics. So sit back, relax, and let's proceed with one of the most powerful tools in your manufacturing tool kit. Statistics will help you center your processes well within the specification limits.

In the last chapter you learned that upper and lower control limits differ from specification limits. A control limit is a statistical term that refers to the limits of the variability of the process. Statistics state the variability of a process in numbers. Through the use of some clever mathematics, it is possible to predict how much a process will vary over time. Again, this is like the car driving down the freeway. Control limits define an imaginary lane in which you will be driving 99.7 percent of the time. In a very broad sense, the control limits predict the future—not the future for any single event or moment in time, but rather the future trend of groups of measurements and values.

There are more powerful tools in statistics than just control limits. Statistics can define more accurately the *spread* of the values in which future readings will fall, assuming that the process is under control. This spread can be utilized to understand and define the capability of the process. Your goal is to improve the capability of your manufacturing processes, and by exploring some of the more powerful tools of statistics, you can achieve that goal. First you must understand how to measure this process spread.

THE NORMAL DISTRIBUTION

Nature causes variation in manufacturing. In fact, this variation is so constant that it actually has an identifiable shape called a *normal distribution*. See if you can find the shape of natural variation in the next example of tire pressure at Everest. The employees wanted to know how much variability they could expect in tire pressure. They used a *histogram* type of chart to analyze the data.

A histogram represents a variable on the horizontal axis—in this case, tire pressure readings—and it plots the frequency

FIGURE 10 1

Simple Histogram with Normal Distribution

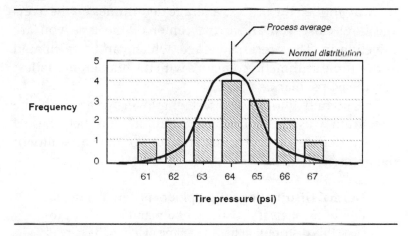

Tire pressure (psi)

of that value along the vertical axis. This type of chart shows the shape—symmetrical around a center value—of natural variation in a process. A natural shape tapers off as the values move farther away from the center point. The histogram shown in Figure 10.1 above says that it is more likely for a tire pressure value to be 64 psi than 61 psi or 67 psi.

On the histogram, the Everest employees placed two lines to evaluate their data. The first line is the *process average* or *mean*, and is illustrated by the vertical line in the center of the histogram. This vertical line represents the average value, or the average of all the tire pressure measurements taken. It is also the value that future readings will fall at or around.

> **Process Average or Mean**   Arithmetic center of a grouping of process measurements calculated by adding the measurements and dividing by the number of measurements taken. *Other terms—x bar, arithmetic mean.*

The process average is an important concept because it locates the center of your process window. If you think of the freeway model, the process average determines on the average how close you are to the center of the lane you are traveling in. The process average, which can be calculated with simple arithmetic, coincides with the histogram's tallest bar, the 64 psi mark.

The second line the Everest employees placed on the histogram was the normal distribution. This bell-shaped curve describes the theoretical measure of the variability of the process.

> **Normal Distribution**   A graphic or pictorial presentation of the *normal* variability of a manufacturing process, this distribution has the shape of a bell. Its spread and center location are calculated with statistical analysis of the process measurements. *Other terms—bell curve, standard distribution.*

The normal distribution curve is calculated with a formula that compares each reading to the process average. The more the readings are *spread* away from the process average, the wider the curve will be. In fact, the curve actually defines the spread and is an accurate representation of process variability. A curve with a small width represents little variability; a broad width has a lot of variability. If the original measurements were accurate, it is highly likely that future values will fall within this curve. It is important to keep in mind that the curve cannot predict any particular pressure in the future, but rather the range and the spread of future tire pressures that are displaying natural variation.

Individual data may vary slightly from measurement to measurement without affecting the shape of this curve. This is an important point. We are only interested in a change to the process—for the better or for the worse. We are not interested in an individual value.

The curve, however, is sensitive to process changes since it will change its shape or location if a process change does occur. The curve is sensitive to a change in the process and not sensitive to one data point. This is exactly what we were looking for—a tool that describes the variability of a process over time. Reading and understanding these curves is similar to reading trends in control charts. The *spread* of the normal distribution and its location provide knowledge and understanding of the process. Generally a tall, thin curve means less variability. A longer, flatter curve means greater variability and a wider process spread.

Process Spread

Through process control, you can reduce the variability of a process. This was demonstrated in Chapter 9. Statistics state that if you reduce the variability in the process, the shape of the normal curve will change. The improved process should produce a curve that is taller and thinner, which would indicate a tighter spread of values.

Everest employees tested this theory by plotting Tuesday's and Wednesday's data on tire pressures to see if the curves were any different. You can see in Figure 10.2 that the taller curve that represents Tuesday's process has a smaller spread that goes from 62 psi to 66 psi, while the Wednesday curve runs from 61 psi to 67 psi. That's quite a difference. The taller curve represents a process that is more in control.

This proves one theory of statistics. The less variability in the manufacturing process, the taller and thinner the normal distribution. You are hitting your target more often. Tall thin

FIGURE 10.2

Improvement in Variability

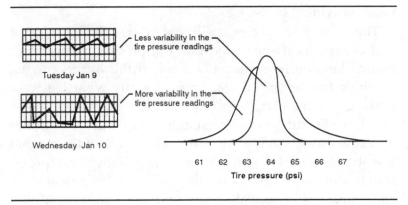

curves reflect a good process. Your goal is to reduce the process spread. See Figure 10.2 above.

Statistics provide a mathematical method of measuring variability and a measure of your process center. This is important for good process control. You can see the variability, and you can also measure the average amount of variability. Statistics provide a number—called a sigma unit—that describes the shape of the curve.

> **Sigma** Numerical measure of the variability in a product or process determined by statistical sampling of a population and designated in formulas by the Greek letter sigma(σ). The formula is the square root of the sum of the mean squared average difference between each measurement (whew!) and the process average. *Other term—standard deviation.*

A normal distribution curve is divided into six *sigma* or *standard deviation* sections (which make up 99.7 percent of the curve). A sigma unit describes variability—just as the normal distribution describes the total variability. The sigma or standard deviation is basically an average of the variability around the process average or mean. It is calculated with a formula that analyzes a sampling of values from the process and compares each value to the average value for the group. Generally speaking, the lower the sigma value, the lower the variability of the product. We are seeking low standard deviation numbers, since they signify a process with less variability. The sigma value and its associated normal distribution curve are universal in nature because they are a picture of natural variation of populations.

Everything belongs to some population group. All people, all products, or in this specific case, all bicycle tires produced at Everest, are members of a population. Natural variation exists within all population groups. The normal distribution curve reflects this variability, and 99.7 percent of this variation will fall under the curve. We predict that all new members of the population will fall within the normal distribution. However, when uncontrolled variation enters the process, the new measurements will fall farther and farther from the process average or mean. Factors other than natural variation are influencing the process.

Generally, you can't test the entire population to determine the normal distribution. You select an appropriate sample, and test the variable characteristic. You generalize from the data that they are representative of the entire population, and you plot the normal distribution. You analyze how new measurements compare with your representative sample.

The resulting curve (see Figure 10.3) represents the variability of characteristics that exist extensively in nature and in manufacturing. How is the curve drawn over a histogram? In the case of tire pressure, data from tire pressure readings are accumulated over a long period of time. This is the sample

FIGURE 10.3

Normal Distribution Curve

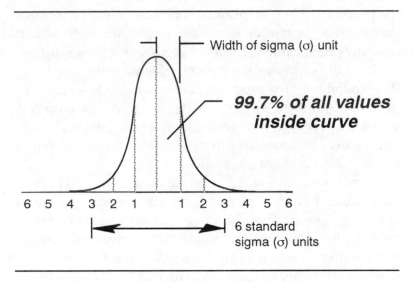

that represents the entire population of bicycle tire pressure at Everest. The mean, or process average, and sigma are calculated. Based on the data we have been using, the mean is 64 psi, and the sigma or standard deviation is 1.07 psi. Each segment on the chart represents 1.07 psi of variability around the process average. The variability of the process is 6.4 psi. The curve constructed for this tire pressure example can be seen in Figure 10.4 on page 143.

You can now measure improvement over time. When you reduce the variability to 6.0 and 5.7, you can monitor your progress. The higher the standard deviation—or sigma—the wider the process spread. The lower the number, the thinner the process spread. You want a low sigma.

FIGURE 10.4

Normal Distribution Curve

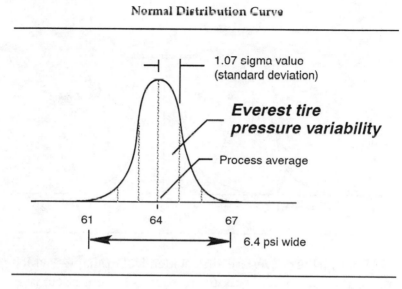

Process Centering

Reducing variation in a process is your goal because it increases the likelihood of good product. However, normal distribution does not tell the whole story. Remember that the numbers on Tuesday, the day Everest started to go out of control with the seven values above the center line, were uniform and did not have much variation. The normal distribution curve would have been tall and narrow. However, the process was tending out of control. The process average for Monday was 65.07 psi, up slightly from the previous average. The process average for Monday afternoon was 68 psi, which wandered away from the original process average. If you chart the normal distribution curves, they would be almost identical, but would show a significant shift to the right. See Figure 10.5.

FIGURE 10.5

Process Center Shift

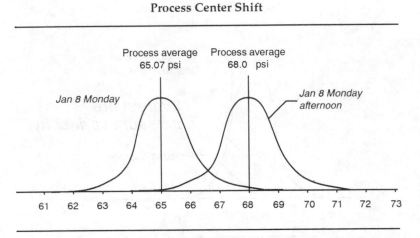

Process average 65.07 psi   Process average 68.0 psi

Jan 8 Monday   Jan 8 Monday afternoon

61 62 63 64 65 66 67 68 69 70 71 72 73

Although the curves are almost identical in shape—which means the variability is about the same—the entire curve is starting to move. How do you know when uncontrolled variation is at work, and how do you know that a change is taking place? The answer is actually simple—you compare the process average to the specifications. If you overlay the specification limits on the normal curves, it becomes obvious when you are trending outside the specification limits.

When the two curves are placed within the specification limits, you get a clear picture of the trend. Monday afternoon's curve is moving toward the upper specification limit. Nothing is out of specification yet, but the curve reflects a potential problem on Tuesday. Tuesday's process has product outside the specification limit. Notice the short, fat curve, and notice that the process average is much higher than the previous 65 psi. It also tells you that any additional product will probably be outside the specification limit. See Figure 10.6.

Defective product and scrap could have been avoided by investigating the trend on Monday and taking corrective

FIGURE 10.6

Process Centering in Relation to Specification Limits

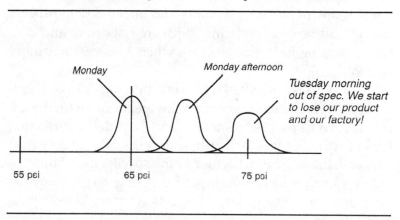

action. The change in the process average was an action signal. You want to keep the process centered well within the specification limits. In this manner, you have an excellent chance of making it right the first time. This is accomplished with a valuable tool termed *process capability index*, or *Cpk*.

Capability Performance Index (Cpk) Compares the variability of a process with the factory or customer specifications and reveals how centered the process is within specifications. *See capability potential.*

Basically, the Cpk is a ratio of the specification limits to the width of the normal distribution curve. The following is a formula for Cpk calculation:

$$Cpk = \frac{(Upper\ spec\ limit - Lower\ spec\ limit)}{6\ standard\ deviations\ (sigma)} \times \left(1 - \frac{|Target\ value - Process\ mean|}{Spec\ width/2}\right)$$

This formula combines the process spread (six standard deviations) and the centering of the normal distribution. Cpk is a measure of process capability and includes both process spread and process centering. Both are important and contribute to a higher value for Cpk when a process has more control.

A high Cpk is associated with a tall, thin standard distribution curve that is centered within the specification limit. Cpk is sensitive to process centering and includes a correction factor to penalize you for drifting away from the target value. Target value is the exact center of the specification limits. In the example we've been using, 65 psi is the target value because it is the center of the specification range, 55 to 75 psi.

If the normal distribution is centered, and the curve fits entirely within the specification limits, the Cpk will have a minimum value of 1.0. If even a small portion of the curve goes outside the goal posts, you will go below 1.0. This 1.0 value is a starting point for Cpk values that represent a process that is under control. Continuous improvement in a manufacturing process will move this value higher and leave some margin for error or drift. If the number is somewhat above 1.0, the chances of the process exceeding the specification limits will be minimized.

For this reason, and for some statistical reasons (that would make good outside reading), many manufacturers have adopted a first step standard for Cpk value of 1.33 minimum. A 1.33 Cpk allows some process drift and a small margin of error before trouble begins.

The Cpk charts in Figure 10.7 illustrate several normal distributions for tire pressure at Everest during a month's operation. Notice that a reasonable Cpk value of 1.33 can only be obtained if both process centering and process spread are kept in control in relation to the specification limits.

FIGURE 10.7

Cpk Charts for Tire Pressure

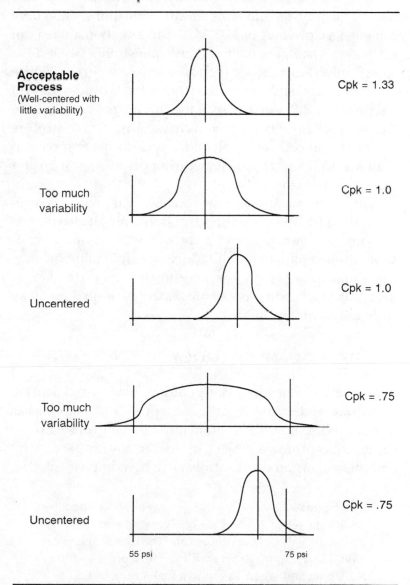

The top curve in Figure 10.7 is the most desirable shape because it minimizes the variability of the process and also centers the process within the specification limits. This curve achieved a Cpk value of 1.33. With this situation, there is an extremely small probability of any value going beyond the specification limits. Study the additional four curves. Notice that a normal distribution curve that is too fat and wide will have a lower Cpk value. Also, if the curve is not centered, the Cpk drops. Remember that both characteristics of control are essential—process centering and a small process spread. If both are achieved, the manufacturing process has a greater likelihood of high yields.

A high Cpk not only assures you that your process is *capable* of running well within specification limits, it also detects uncontrolled variation and improvements in the process. Uncontrolled variation will reduce the Cpk value quickly, signalling the need for action. Investigate and find the assignable cause. Cpk monitors any movement of the process away from the natural variation.

SIX SIGMA & MOTOROLA

One American company, Motorola, has done an excellent job of defining and progressing the concept of Cpk. Motorola has an outstanding internal program for conveying to its employees the concept of *Six Sigma*, a Cpk goal achieved after care-ful and meticulous process controls have been implemented.

> **Six Sigma** Process control program developed by Motorola sets a goal of a six sigma spread between the target value of the process and the closest specification. The Six Sigma Program allows for a process shift of 1.5 sigma and is equivalent to a Cpk rating of 1.5.

Six sigma is an important part of the Motorola strategy because the proper use of statistical concepts can provide a foundation for high quality products with minimal problems. Motorola knows that if it achieves a six sigma process everywhere within its plants and with its suppliers, the chance of defective products is extremely small. With the six sigma approach, the chance of a defect is only 3.41 parts per million. Another way of thinking of 3.41 parts per million is to convert that number to yield. This yield equates to 99.999660 percent. That's pretty impressive, isn't it? How do they come up with that number? Let's explore six sigma.

Using the sigma measure as a guide, Motorola states that there must be six sigma units between the target value (exact center of the specification limits) and the process specification limit. These six sigma units provide the safety margin previously discussed with Cpks that are above 1.0. In fact, the Motorola approach gives an equivalent Cpk value that is above 1.5 while still allowing the process center to drift 1.5 sigma in either direction from the process target value. Motorola understands that there is variability in every process and has developed a comprehensive program of statistical process control. The company strives for and achieves minimal variability and processes that are centered within the specification limits. See Figure 10.8 on page 150.

Even if a process shift of 1.5 sigma occurs, there is significant safety room on either side of the normal distribution curve (the extended comfortable safety margin of 4.5 sigma) with the six sigma approach. It is very unlikely that any part or process will fall out of specification. How exactly does Motorola achieve a tall, thin normal distribution that fits between the specification limits with plenty of margin and still allows a process shift of 1.5 sigma? The Motorola program has been highly successful because the company

FIGURE 10.8

Motorola Six Sigma Program

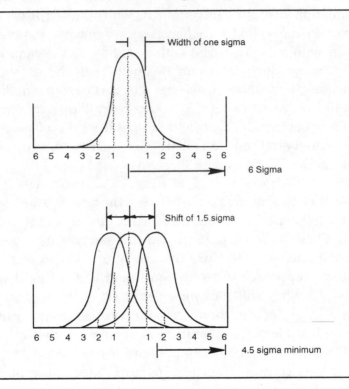

involves its operators and emphasizes continuous improvement toward reduced variability through several techniques. It utilizes fishbone diagrams, P charting, control charting, and design of experiments.

Design of Experiments

Even with the new statistical tool of Cpk, Everest ran into some trouble with its process. Everest wanted to achieve a

Cpk of 1.5—a value that would assure 3.41 ppm defective or less. The best value the company could obtain for a tire pressure Cpk was 1.2. It could not get the process up to a 1.5 Cpk. In order to do that, the Wheel Assembly department would have to improve its ability to fill a tire with air so that it could achieve a centered normal distribution curve. The team tried everything—fixing the pump, making the instruments more accurate, and working with the control chart—but they were at a loss. Then they discovered *design of experiments*—a one-time test of the process. If the test is set up correctly, it is possible to learn the key parameters that will produce a more uniform and repeatable process. Cpk is still very important, but the design of experiments emphasizes analyzing and planning the process so it has a chance of running better the first time. Everest ran a series of tests to determine the best way to fill a tire. Its test plan varied the pump psi, the release valve types, and tires from different vendors. These are important variables to study because each has a different characteristic. Variables were tried in different combinations to determine which combination worked best. The following table illustrates the various tests:

Test	Pump (psi)	Valve	Tire Type	Cpk
1	65	Acme	Everlast	1.03
2	70	**Acme**	**Everlast**	**0.99**
3	65	Best	Everlast	1.01
4	70	Best	Everlast	1.07
5	65	Acme	Knob Tires, Inc.	1.60
6	70	Acme	Knob Tires, Inc.	1.50
7	65	Best	Knob Tires, Inc.	1.52
8	70	Best	Knob Tires, Inc.	1.63

In a comparative study, a *control* is needed to compare data. The control in the above case is test #2, which is Everest's current method of operation. A control serves as a reference point to note improvement or decline in the process. Prior to running the test, the gages were calibrated to assure that extra variation didn't influence the test. Cpks were calculated for each test, and each Cpk was listed in the right-hand column.

There are many scientific ways to analyze the data, but it is obvious that Knob tires yielded a consistently higher Cpk than Everlast. The Cpk readings on tests 5, 6, 7, and 8 were considerably higher than tests 1, 2, 3, and 4. The one item in common with 5, 6, 7, and 8 is that all four tests used Knob tires rather than Everlast tires. Since Knob tires yielded the highest Cpks, this combination will probably improve Everest's process. If Everest switches to Knob tires, a Cpk value of 1.5 may be achieved. This possibility probably wouldn't have been discovered if the Everest team had relied solely on process control for process improvement. Design of experiment is a valuable tool.

> **Design of Experiment (DOE)**   This formal method of troubleshooting and problem solving utilizes a controlled experiment to discover the important variables in a product or a design. A change in these critical variables may significantly reduce variation.

DOE reduces variation and improves process centering. Although certain types of DOE use mathematics for analysis, operators can perform simple DOE experiments with a good working knowledge of the manufacturing process. When you analyze and improve manufacturing processes through the use of statistics, you are engaging in a sophisticated technique called *statistical operator control*. At this point, you are indeed working smart.

Statistical Operator Control (SOC) The operator monitors the variability of the process by charting and feedback loops, and performs process corrective actions when excessive variability occurs.

With SPC, SOC, and DOE, (come on, you know what they mean) you can gain control of your work area, and improve your process and product by minimizing variability and centering your process well within the specification window.

TIPS

(1) High Cpks provide a safety margin between the process and the specification limit. Use your skills in variability reduction and process centering to continually improve the Cpk values in your area.

(2) Strive to center your process inside the specification limits. Centering reduces variability and improves your manufacturing process.

(3) Watch your Cpks for trends. A sudden movement downward may indicate uncontrolled variation. Look for the cause. Movement upward may indicate progress in your attempts to improve your process. It provides valuable feedback for your efforts.

(4) Design of experiment, when properly performed, can greatly reduce the effort required for process control and quite often provide a higher degree of process capability than can be obtained by SPC alone.

(5) Use a control group that represents your present process in your design of experiments. It is your best reference point.

CHAPTER REVIEW

(1) Consider two processes. Process A for tire pressure has a normal distribution with a spread of 6 psi, and Process B has a spread of 6.5 psi. Which would be the better process? Which process would have the taller normal distribution curve?

(2) If high Cpks are important for safety margins, why wouldn't a manufacturing company move its specification limits out far enough that the process always has a safety margin to it? In the case of the wheel assembly at Everest, why not move the specification limits to 45 psi and 85 psi? Explain your answer.

(3) Can you have a process that produces bad product with low variability? Explain.

(4) The following Cpks were recorded on Everest's tire pressure—1.1, 1.2, 1.05, 0.3, 1.2, 1.2, 1.1. Which group of readings would indicate uncontrolled variation? Should the process be stopped with the above readings? Is there a probability of defective product being manufactured?

(5) To achieve a Cpk of 1.0, which of the following are the correct sigma units between the process target and the specification limit—two sigma, three sigma, or four sigma?

(6) The design of experiment indicated that a switch to Knob tires would improve the capability of Everest's manufacturing process. Since Everlast tires had been a valued supplier up to this time, Everest decided to share the DOE technique with Everlast to see if that company could improve the quality of tires enough

to meet or exceed the performance of Knob tires. Do you think the supplier could benefit from DOE, or should Everest just switch to Knob tires? Explain your answer.

(7) Can you think of a place in your operation to use the Cpk technique for process capability? List the location and a plan of action.

(8) Everest calculated more sigma values on tire pressure for the last two weeks in January. Which of the following sigma numbers represent the least variability in tire pressure—1.08, 1.10, 0.71, 1.31, 1.25 psi?

Chapter Eleven

Get to the Root of the Problem!

Rule Eleven

Fix the problem, not the blame.

T he ultimate goal in manufacturing is to produce a defect-free product every time. Essentially that means trying to produce identical, perfect products—which, as you learned earlier, is statistically impossible. Nature and human tendencies constantly work against the goals of manufacturing. This presents a challenge. When defective product gets through, you can't point fingers at Mother Nature or human error. Customers want good product, not excuses. You have to find the solutions to the problems, and find ways to overcome these natural obstacles.

We are all imperfect beings trying to create perfect product. Human error can be traced to the root or cause of most manufacturing problems. What does it mean to get to the root of the problem? Frequently, it seems to mean that you find the person responsible for the problem and fix the blame. At least as much time is spent blaming suppliers, coworkers, and managers as is spent fixing the problem.

When a problem arises, it can almost always be traced back to human error. In fact, you can usually follow the trail back to the person who made the error. When the person who made the mistake is found, there is often a tendency to focus more on his or her shortcomings than on the problem. Sometimes the person is viewed as clumsy, careless, or stupid. Rather than faulting the design, the material, or the process, the person is singled out. The blame is often fixed before there is an attempt to fix the problem. Valuable time is wasted fixing the blame, and the people involved become emotional wrecks in the process.

What is generally the first reaction when a child spills his or her milk? It's probably something like, "oh no, you spilled milk all over the table." There is a tendency to fix the blame before you even grab the towel. The adult gets angry, and the child starts crying. Meanwhile the milk is running off the table onto the floor. Whether or not you get mad at the child, and fix the blame, you still have to correct the problem. Emotional reactions just complicate the situation and add nothing to the solution. The milk still must be cleaned up.

An ingenious device was invented because all children have a tendency to spill milk. This container with a spill-proof lid eliminates the problem and the need to fix the blame. If you focus on ingenious devices in manufacturing that compensate for human tendencies, the manufacturing environment then becomes *user friendly.*

Let's face it, no one wants the blame to rest on them. When we make a mistake, most of us feel that there is something wrong with us. Instead of focusing on the positive and viewing a mistake as an opportunity to improve knowledge and skills, we just feel badly about our weaknesses. When people around us share this opinion, it tends to confirm that we are flawed. Rather than accept the blame for a problem, we often try to cover up, or *pass the buck*. Again, we are fixing the blame, not the problem. Learn to fix the problem.

In an earlier chapter, when the lock and key design was discussed, an example was cited of the Everest employee who inserted the bearing into the housing upside down. Most workers would probably insert the bearing upside down occasionally. An ingenious device was invented with the lock and key design that prevented this problem from occurring.

Before the flange was attached to the bearing, it was the worker's responsibility to insert the bearing correctly. Since this was his responsibility, and he knew that he was prone to making this mistake, he should have checked his work carefully each time to ensure that he didn't repeat his error. The point is that there may be inherent flaws in the design, material, or process. These flaws may make you prone to mistakes, but you are always responsible for your actions. When you recognize your tendency to make certain mistakes, it is your responsibility to prevent them.

Frequently, the design or process must be reworked or a new material found to reduce or eliminate certain problems. Sometimes, though, the problem rests squarely with the worker. If you continually make the same mistakes because you are not following procedure, you should reexamine your goals. Are you interested in producing quality product, or are you only interested in your paycheck? If the paycheck is your only bottom line, you are cheating yourself and the company.

Other jobs may suit you better, and other people may be better suited to your job. If producing quality product is important to you, follow the suggestions made earlier in the book:

- Get formal training.
- Continually educate yourself.
- Keep daily logs.
- Communicate and participate.
- Perform daily prevention checks.
- Ask questions, and seek answers.
- Question your data.
- Organize your work area and paperwork.
- Seek continuous improvement in all areas.

The more educated and skilled you become at problem solving, the less you'll have to rely on spill-proof containers!

CHALLENGE YOUR DATA!

Communication within a factory can be difficult. Factories are very noisy and filled with activity. Work is moving from station to station; machines are operating; and people are walking around and talking. With all this activity, basic communication requires an extra effort. Decisions are made based on facts, figures, and communications that are received in a hurried, noisy environment. All the distractions make it difficult to analyze the information.

Your success depends on accurate information. Make sure that you understand the information before you act upon it. You will naturally come to rely on several sources of data in the factory. However, you should periodically verify that the data are correct. If all employees in a factory practice this procedure, no one will be offended when their data are occasionally checked for accuracy and authenticity.

A common source of misinformation in a factory is the computer. People generally *trust* the information in computers and rarely challenge it. Computers are only as accurate as the data fed into them. If the original data were incorrect, or if the operator made an error when inputting data, the computer will display incorrect information. This problem is so widespread that computer experts have a term for it.

> **Garbage In/Garbage Out (GIGO)**   The term commonly used to describe the problem of entering bad data into a computer database, thereby causing the generation of false reports.

Be cautious about computer data, and constantly check it for accuracy. Computers are not the only source of misinformation—it may also come from your coworkers, or your measuring devices may not be calibrated correctly. Rick in the Paint department at Everest learned this lesson the hard way.

Remember when in an earlier chapter Rick tried to cure the paint faster by increasing the oven temperature? Rick recognized his mistake and corrected his actions. Unfortunately, two months later, bikes started returning to the factory with peeling paint. Everyone looked at Rick as the likely cause of the problem. Rick was certain that he was now following procedure, but since the problem was in his area, he felt responsible. Rick wanted to know why this problem had recurred, and he started to investigate.

Rick reviewed the documents that traveled with the bikes as they progressed through the shop. These documents verified that the faulty bikes were painted on March 16—several weeks after Rick recognized and corrected his mistake. However, when Rick checked the documents during that period of time, his control chart for March 16 was missing. At that point, some people thought that Rick hadn't checked the oven temperature on March 16, and that he was deliberately trying to cover his mistake. Rick kept up his search—why were the documents missing?

Rick's control chart for March 15 was also missing. Rick checked on a calendar and discovered that March 15 and 16 were a Saturday and Sunday. He didn't work weekends. Shipping had entered the wrong date into the computer. Rick checked his control chart for March 14 and verified that his temperature setting was correct. This information didn't make sense. Why? He studied the document further and discovered a written notation on the chart indicating that maintenance had calibrated the oven the night before. Rick immediately talked to Maintenance about the procedure they followed to calibrate the oven. Apparently they had put the temperature probe next to the seal on the door where the temperatures are cooler. This is an incorrect position for a calibration probe, and the Maintenance department was not following procedures that clearly stated the probe was to be placed closer to the heating element where the product would normally be positioned during the baking operation.

Since Rick put the bikes farther back in the oven, closer to the heating element, this resulted in a slightly higher baking temperature in the area of the bikes. The problem was solved. Maintenance learned to place the probes correctly, and Rick learned to question all sources of data. The lesson here is one of fixing the problem, not the blame. Finding the cause is not the same as finding the blame. When you want to find the cause, you ask, "why would this happen?" Look for a solution and a way to prevent the problem from happening in the future. When you want to fix the blame, ask, "who did it?" The *who* is not as significant as the *why*.

Answering why leads you in the direction of prevention; who just confirms that humans have a tendency to make mistakes. Learn to ask the question why to the bottom of a problem. If Rick hadn't asked why at least three times, the problem with the cracking paint might still not be solved.

PROBLEM SOLVING

Problem solving skills are a valuable asset, but are somewhat difficult to acquire. The key to problem solving is an inquiring mind. You play detective; knowledge leads to the solution. The following is a common-sense program to help you with problem solving at your manufacturing facility:

(1) *Define the problem.* Start with a clear understanding of the problem. Is the scrap too high? Are the parts out of tolerance? Is there not enough capacity? Many problem solving attempts end in frustration be-cause the problem is not clearly defined. In order to define the problem, you must define your objective. You want to reduce scrap by 20 percent. The scrap is 20 percent higher than it should be. When you reach your objective, the problem is solved.

(2) *Communicate the problem.* Inform your supervisors and managers of the problem, and ask for direction. Offer suggestions regarding possible solutions. If you are a supervisor, offer clear direction and assignments to those who will be given the task of working on the problem. Frequently, when the problem is properly communicated, someone in

the organization will recognize the problem and have the knowledge to provide a quick fix.

(3) *Contain the problem.* If the problem is related to quality, as many problems are, it is imperative that the material, components, work-in-process, and all other inventories be located and quarantined if necessary. The questionable process should be responsibly shut down until the problem is solved.

(4) *Challenge all data.* Never assume. All data are suspect when you get into a problem situation. Systematically check all instrument data, computer data, verbal reports, and paperwork for accuracy and consistency. This will save valuable time and may save you from going down dead-end paths.

(5) *Involve coworkers and suppliers in the problem.* Two heads are better than one. Involving more peo-ple in a problem, especially your suppliers, may lead to a faster solution. Good suppliers have knowledge of their product and how it is best utilized in different applications. Also, it is always possible that your problem started outside your company at a supplier. Your suppliers will want to know if they are delivering defective product.

(6) *Find the root cause.* This is the most difficult part of problem solving. Keep penetrating the problem and asking questions. Divide the problem into its design, material, and process components. Then proceed to analyze the interrelationship among the design, the material, and the process. Somewhere within the three lies the root cause. Remember that in most cases, human error is responsible for problems in

manufacturing. You are not looking for the blame, but for the system problem that can be fixed.

(7) *Fix the problem, not the blame.* Find and fix the causes with a solution that yields the most lasting effects. The design, the material, and/or the process may be the cause of the problem. Your job is to define the problem, find the cause, and perform the corrective action that will improve the process. If you perform these functions as a team, and forget about blame, the best will surface in everyone. Everyone's common goal is a quality product. The rest is small stuff, and of course, you don't sweat the small stuff!

After identifying a solution, thoroughly test and communicate your solution. Make sure it really works and doesn't cause another problem later in the process and result in telegraphing. Also verify that your solution follows specifications and procedures. If not, communicate this issue so that the specifications and procedures can be reviewed. Finally use one or several of the 12 frontline rules to permanently fix the problem, and prevent it from recurring. Remember, you are part of a team and should approach problem solving as a team.

Every Problem is an Opportunity

In summary, a manufacturing facility will have its problems. Some of the greatest advances in your factory will probably come from resolving problems, frustrations, and workmanship issues. A truly wise person or organization can learn from the mistakes of others or from past mistakes. You don't have to make the same mistakes over and over. You can learn, and move forward. When a design creates problems, modify the design, and thoroughly test it out. It then becomes a better design. This is another part of continuous improvement. Eliminating problems means that you are getting better and better with time. Your product is becoming more perfect. When a problem arises, and you recognize it as a problem, you have the opportunity to improve. Eventually you can become so good at this that you will be able to determine where potential problems lie. You will be able to make corrections *before* the problem actually materializes.

One of the secrets to success is understanding that opportunity lies hidden in each problem. If you and your coworkers can deal with problems, and learn to be flexible, intelligent, and inquisitive, you will be part of a dynamic, progressive team. Problem solving can become a significant portion of your continuous improvement program within the factory. With each solution, you can improve your quality ethic and your pride of workmanship.

TIPS

(1) Divide a problem into its design, material, and process components for an analysis of the possible root causes and solutions. Test each system separately to determine its effects.

(2) Use your supplier's internal engineering and research systems to find solutions to difficult problems.

(3) Most computer errors occur when entering data into the computer. Examine this area closely to assure that the data entry process is accurate.

(4) People are more cooperative when their knowledge and skills are respected. Involve your suppliers and coworkers in your problem by making them a valuable part of the problem solving team.

(5) Beware of solutions to problems that can create other problems. Carefully test your solution both in terms of your area and how it affects later operations.

CHAPTER REVIEW

(1) Perform the following test with your coworkers. Draw six overlapping geometric shapes on a piece of paper, and cover up your work. Carefully describe what you have drawn to your coworkers. Ask them to draw the figure based on the information that you describe. How did they do? Do you see how difficult it is to communicate accurate information?

(2) How many different sources give you the information you need to perform your job? How would you verify the accuracy of the information?

(3) Occasionally a problem can be traced to inadequate worker training. An analysis of which system, design, material, or process reveals the underlying root cause? When the problem was traced to the workers inserting the bearing into the housing incorrectly, which system or systems required correction? What is the bottom line root cause of most problems in the factory? How can you correct this?

Chapter Twelve
Personal Improvement

Rule Twelve

Continuously improve your personal skills through a positive attitude, team involvement, communication, and education.

Good news is on the horizon for the American worker. More and more companies view the line worker as an asset. Your knowledge and ideas are valuable resources. Five years ago this book would not have been considered marketable. Workers weren't viewed as solutions to the problems plaguing American manufacturing; they were generally considered *the problems*. Upper management was thought to be the valuable human resource while line workers were considered interchangeable parts. The trend today is to view line workers and managers as a partnership—one cannot reach the goals without the help of the other. Good, competitive products are possible when everyone shares and works toward the same goals.

Worker involvement makes the job more interesting and meaningful, but it also puts more responsibility on your shoulders. This responsibility carries with it the constant need for self-improvement. Your company can encourage and motivate, but the ultimate commitment rests with you. You have to take the active role. Reading this book indicates that you recognize the importance of improvement and learning. Personal improvement programs keep you marketable and competitive.

In the 1989 Universal Studios release, *Dad,* Jack Lemmon's character states that, "in America, anything is possible if you show up for work." Effort produces results. Anything is possible if you put forth the effort, but you have to decide what things are worth the effort. If you put little into your day, you will get little out. If you make it an exciting learning experience, you will reap the gains. There are many self-help books on the market today. Each has the same basic message: If you want to change or improve something in your life, *you* must put forth the effort. Life owes you nothing except the right to reap the benefits of your efforts.

Effort leads to improvement, and lack of effort leads to degeneration and atrophy. When people lie around and *veg,* bodies *and* minds get fat and flabby. A precious gift is lost— the ability to learn and analyze new information. Just as our bodies need good food and exercise, our minds need exercise to stay fit.

In a television commercial, Cher tells us that we are more concerned with what we put into our cars than what we put into our bodies. This can be taken one step farther—we are more concerned with what we put into our bodies than what we put into our minds. In an increasingly competitive world, you need to feed your mind stimulating information to keep up. This need doesn't end when you complete your formal education. Formal education is really just the beginning. You need to find new and exciting ways to apply the information. Deming's simple formula for quality product—plan, do, study, and act—can also contribute to a quality life.

SET YOUR GOAL POSTS

Goals give you direction and focus. Decide what you want to achieve, and set your goals. What actions will benefit your personal growth and career? Outline the steps that will move you closer to your goals. Make a plan. Use the notebook (that we're sure you've started by now) to list your goals. Set goals for the day, the week, the month, and the year. Generally, the daily goals will be stepping stones toward the weekly goals. The weekly goals will be stepping stones for the monthly goals, and so on. Deadlines have a tendency to get us moving. Set specific deadlines for each goal.

Plans are relatively easy to make. *Doing* is the hard part. You may not actually think about it, but you determine how much your goals are worth to you by how much effort you put forth to achieve them. Very little effort is put forth to achieve things that aren't worth much to you, while a great deal of effort is put forth to achieve those things that have worth. When you examine the goals that you never seem to accomplish, you can learn a great deal about yourself. Sometimes you learn that you really don't want those things after all. Let those things go, and pursue other paths. Other times, you learn that you need to push yourself harder. The important thing is doing.

The way to tomorrow is paved through today. Today's accomplishments are tomorrow's foundations. If you don't accomplish today, it becomes less likely that you will accomplish tomorrow. The more you do; the more you *can* do. It's a magical formula. You develop discipline by moving toward goals.

The next step is to study your progress, and correct your actions. Sometimes you won't reach your goals, even when you've put forth your best effort. Many people give up at this point—calling themselves failures and figuring that they don't have what it takes. This so-called failure is just an indicator that there are additional steps to reaching the goal; there was a miscalculation at the beginning. The *failure* shows the weakness that must be corrected—and it moves you one step closer to your goals. The next step is corrective action. Learn from your mistakes, and forge ahead. Reaching goals may not be easy, but they form the stairway to success.

The following are basic goals that may improve your personal and professional life.

Communication Goals
The ability to communicate both verbally and in writing is an asset for most jobs. There are many excellent books and classes available that will improve these skills. Writing and speaking clearly and concisely require practice. These skills increase the likelihood of your message being heard and understood. They also increase the likelihood of your understanding other people's messages.

Team Goals
Energy and ideas form a powerful force. When people work together, the whole is greater than the sum of its parts. Each person has a unique perspective, and many different perspectives result in a larger picture. If you are asked to participate, do it! If you are not asked, volunteer! Teamwork is an exciting

learning experience requiring you to listen to and understand a variety of different *mental positions*, and contribute in a positive manner. Teamwork improves your analytical thinking and problem solving skills.

Idea and Action Goals

The more ideas you generate and act upon, the more ideas you will generate. It's that magical formula, again. If you open the door to your mind, it will be like a never-ending well. The key is to act upon your ideas and more will follow. So many ideas will follow that you will have to learn to focus and channel. Your notebook will become a constant companion and a storehouse for all your ideas. Nurture your ideas, and act upon them. Companies welcome new ideas that keep their product fresh and cut costs.

Educational Goals

Read. It is the best source for self-improvement. Take classes, and attend seminars. Attend all training offered at work. Technology changes rapidly—if you don't keep up, you'll be left behind.

Positive Focus Goals

Focus on those aspects of your life that benefit or lead to your goals. Find the positive, and focus on that. Most things and people are not inherently good or bad; they're just labeled good or bad. Practice changing the label from bad to good, or resist the temptation to label. Develop a winner's attitude. Every obstacle is an opportunity.

THE WINNER'S ATTITUDE

In *The Psychology of Winning*, Berkeley Books, 1989, Dennis Waitley, an expert on the winning attitude, says, "Winners can tell you where they are going, what they plan to do along

the way, and who will be sharing the adventure with them."
Winners have a plan of action and a positive attitude. If
mopping floors leads to their goal, they gladly mop floors.
They keep their eye on the prize. They don't get caught in the
trap that plagues almost every work environment—the negative focus.

Most people are familiar with chronically negative employees. They complain that other people aren't doing their
jobs, management is incompetent, the raises aren't high
enough, and working conditions are awful. They voice their
negative perspective to anyone who will listen. Frequently,
there is some element of truth in their statements. People
listen, and the negativity spreads like a virus. If you asked
the ones with a negative focus what to do about the problems, they would either be stumped for an answer, or they
would respond that nothing can be done. If these people
were placed in paradise, they would complain that the apples
were too sweet or the clean air too hard to breathe. It takes
effort to avoid the negative focus trap, and it destroys many
potentially good workers. They become part of the problem
instead of the solution.

A bad attitude is hard to disguise. It shows in body language, negative statements, and facial expressions. Employees who are negatively focused indicate an unwillingness to
learn, grow, and be adaptable—all the basics required for a
flexible manufacturing facility.

A good attitude is also hard to disguise. The smiling face,
the warm greetings, the willingness to help and learn are
good indicators that you are working with a winner. The basic
"hello," "good-bye," and "thank you" can go a long way
toward building team spirit and cooperation. If given the
choice, which attitude would you want on your team—the
can do or the *can't be done*? There's no contest. A positive

attitude opens doors and helps solve problems. It positions you on the course of continuous improvement.

Success-Oriented Characteristics

Successful people frequently share personality traits that are sometimes referred to as success-oriented characteristics.

These personality traits fit into work environments successfully and help individuals be promoted into more responsible positions. They include the following:

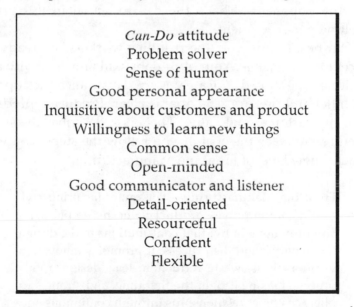

Can-Do attitude
Problem solver
Sense of humor
Good personal appearance
Inquisitive about customers and product
Willingness to learn new things
Common sense
Open-minded
Good communicator and listener
Detail-oriented
Resourceful
Confident
Flexible

The characteristics on this list can be learned. With time and effort, anyone can fit this profile. The list describes intelligent people who care about themselves and others. They welcome new ideas and have a willingness to learn. They view problems as opportunities to move forward, and are just the people American manufacturing needs. How many success-oriented characteristics do you possess? Add the ones you are missing to your list of goals.

FRONTLINE MANUFACTURING

Frontline Manufacturing means that we work together, and we work smart. We are tired of beating down the same old bushes and barking up the same old trees. The old ways simply aren't competitive. Americans are competitive. We won't give up simply because our foreign competitors are worthy opponents. That's the call to action in this great melting pot experiment.

Our best troops, you, the frontline workers, are ready to accept the challenge. Your abundant skill and creativity are waiting to be tapped. You are your best insurance; pick up the gauntlet and prepare. The goal is clear. Produce a product that the customer will value. This book describes the steps required to reach the goal. If you follow the steps, you will reach a new level of Frontline Manufacturing.

> **Frontline Manufacturing** This manufacturing style continually improves product value by emphasizing and promoting individual contributions to the design, the material, and the process. It promotes innovation, value analysis, waste reduction, lean design, robust process, making it right the first time, and strong supplier/customer relationships through continuous education and the participation of all workers.

Take a moment to stand back and enjoy your position. You are an important contributor to your organization and the world economy. You add value to things that people use and enjoy. Take an active role in your life. Define and accomplish your goals. You and your company can become *world class*.

TIPS

(1) You can learn by traveling. Make a point to visit local factories, and study their operations.

(2) Put punch into your personal attitude. A little *can do* adds a lot to any situation.

(3) Over-communicate your ideas; leave nothing to chance.

(4) Smile and say "hello," "good morning," "good afternoon" to your colleagues. Respect them, and they will respect you.

(5) If your company offers cross-training, enroll in it. The more variety in your experience, the better perspective you will have in your work environment, and the better your chances for promotion.

(6) Keep a written record of your personal goals, and check it frequently for progress. Don't be discouraged by failures, view them as opportunities to improve!

(7) Write down some success-oriented characteristics that may work for you. Monitor your progress.

Appendix One
Glossary of Terms and Definitions

Accelerated Life Testing (ALT) An aggressive reliability test to evaluate product performance in a simulated long-term aging environment such as high temperature and/or high humidity.

Acceptable Quality Level (AQL) Older quality term referring to the maximum allowable amount of defective parts for every hundred units. Determined by the amount of defective units within a sample size as compared to AQL charts.

Accounts Payable Amounts of money due to suppliers for goods or services.

Accounts Receivable Amounts of money due to the manufacturer from the customer for goods and services performed or delivered.

Accrue An accounting term for identifying transactions where cash may have not been exchanged.

Active Part A component or part that is currently utilized in scheduled production program.

Adhesive   Glue or cement used to assemble components in a bonding process.

Algorithm   A computer formula for calculation that usually includes several steps or sequences of instructions.

Alpha Error   Error of rejecting a good part that should be accepted. Sometimes referred to as overcorrection. *Antonym: Beta Error.*

American National Standards Institute (ANSI)   Group that promotes standards for manufacturing and measurement.

American Production and Inventory Control Society (APICS)   Active organization that meets regularly, publishes key industry documents, and advances the science of manufacturing and inventory management.

American Society of Quality Control (ASQC)   Key organization for the development of quality systems and education.

American Supplier Institute (ASI)   A nonprofit organization dedicated to quality training in such disciplines as the Taguchi DOE method, FMS, JIT, and QFD. Initially started by the Ford Motor Company.

American Work Ethic   Common belief in the United States that hard work is good and moral by its very nature.

Analysis of Variance (ANOVA)   Traditional mathematical tool for determining the various influences of manufacturing variables. Groups of data are analyzed as different or the same depending on their dispersion characteristics.

Analytical Manufacturing   Type of manufacturing that involves the breaking down of a main component or material into several products. For example, a lumber mill would process a tree into various products such as lumber, sawdust, and mulch.

Andon   Lights used to signal trouble on a machine or a process very similar to a traffic light.

Applications Software   Computer programs that have dedicated functions such as word processing, data base, spreadsheet, or business graphics.

Artificial Intelligence (AI)   Type of software that makes decisions based on a large amount of previous decisions documented by experts in the field and simulates human decision making.

ASCII   Standard computer code representing text and numbers.

Assemble to Order (ATO)   Newer method of manufacturing that allows product to be built with a firm order from the customer and with the specific requirements of the customer—as opposed to building product to general specifications and/or building to inventory with no order. *See Make-To-Order.*

Assembly Manufacturing   Type of manufacturing that fits together components, parts, or materials into a finished product by such techniques as fastening, insertion, bonding, welding, and other methods. *Synonym: Synthetic Manufacturing.*

Asset   All items belonging to a manufacturing firm that can be described to have some sort of value.

Assignable Cause   The cause of uncontrolled variation in a manufacturing process. For example, a power failure may cause scrap and defective product. It is the assignable cause for the scrap and defective product. *Synonym: Special Cause.*

Attribute   An attribute is a quality characteristic of a product that is either present or missing. For example, pinholes in a paint job or the number of reflectors on a bicycle are attributes that are either present or missing. *See Variable.*

Attribute Data   The numbers obtained from qualitative measurements of an attribute. They are counted in a yes/no format and are usually presented in yield charts or nonconforming statistical charts such as P charts.

Audit   A formal examination and review of a manufacturing process or procedure.

Automated Guided Vehicles (AGV)   Carts or containers that automatically follow preset pathways for movement of materials through a factory.

Automated Optical Inspection (AOI)   Machines with video cameras that scan the product being manufactured and send the resulting video signal to a computer for analysis. This analysis determines different characteristics of the part including but not limited to the quality levels.

Automated Process Control (APC)   A method of monitoring and adjusting a manufacturing process with the aid of a local computer microprocessor.

Automatic Storage/Retrieval Systems (AS/RS)   Material handling and storage facilities that use robots or automatic systems to move materials/components in and out of storage.

Autonomation   *See Jidohka.*

Average   *See Process Average.*

B vs. C   Simple design of experiment technique for comparing a current process (C) to a better process (B) by analyzing their respective normal variations.

Backlog   Manufacturing units or value of sales orders that have not been shipped.

Bar Coding   Method of labeling parts in manufacturing with a pattern of wide and narrow lines that can be recognized by a computer scanner for identification and thereby identify the lot and part numbers.

Batch Processing   Type of process manufacturing where more than one part or component is processed together as a group through a work center or work cell. *See Continuous Flow Processing.*

Baud Rate   Corresponds to one bit of computer data transmitted per second and is the measure of the speed of transmission devices between computers such as a modem.

Benchmarking Technique of self-evaluation of a business by comparing all aspects of the business to other organizations considered to be best-in-class. *See Best-In-Class.*

Best-In-Class A factory or product that is considered superior to the competition.

Beta Error Error of accepting a bad part that should be rejected. Sometimes referred to as undercorrection. *Antonym: Alpha Error.*

Beta Site A production testing site for new equipment or processes where at least some actual work-in-process will be evaluated prior to bringing the unit or process on line.

Bill of Lading A receipt given to a manufacturer from the transportation company upon pickup indicating the method of shipment and the ultimate destination.

Bill of Materials (BOM) A complete listing of required parts and materials needed to manufacture a certain product.

Bit Smallest coded piece of computer information possible in a digital computer. Represents either an on or off state. *See Byte.*

Boot Common term for loading the operating system onto a computer when starting up a computer from scratch or resetting a computer.

Brainstorming Technique of problem solving that generates many possible ideas, then selects those ideas that have merit.

Breakeven Accounting term that indicates the balance between costs and revenues, and a point where no profits and no losses will occur.

Broadcast System method of controlling and limiting inventories by using electronic signals to actuate work-in-process or components to move to the next work station or to replenish supplies just utilized. The signal is based on completion of a key assembly or process. *See Golf Ball System.*

Bubble Up   System of alerting sequentially higher decision making personnel in a factory when a problem arises. The system continues the alert until the problem is resolved.

Budget   Forecast of expenses for a given period and for a given work center.

Business Plan   A written document that describes the expected performance of a business for a period covering either the next 12 months or for a specified fiscal year. Strategic business plans are broader in scope and cover the expectations for up to 5 years.

Byte   Unit of coded information for a digital computer that consists of more than one bit of information and usually in the sequence of 8, 16, 32, 64, and so on.

C Chart   A special case of the P chart that tracts the number of defects (as opposed to the number of product units) in an inspection lot and is utilized when a million things can go wrong but only a few do. The defects follow a mathematical relationship called a poisson distribution and this chart requires all samples from the population to be the same size.

Calibration   Determination of the accuracy of an instrument using a reference standard and adjusting the instrument with the appropriate correction factors if required.

Capability Performance Index (CPK)   Statistical term that compares the variability of a process to the factory or customer specifications, but also includes how centered a process is inside the customer's specifications. *See Capability Potential.*

Capability Potential (CP)   Statistical term that compares the variability of a process to the customer's specifications. This is defined as the specification width (S) divided by the the process width (P).

Capability Study   An analysis of a particular process that evaluates the fitness and competence of the process to perform to the factory or product specifications. This is usually expressed in mathematical quantities such as statistical variations.

Capacity Manufacturing term describing the maximum output of a machine, a work center or a factory.

Capacity Requirements Planning (CRP) Planning for a factory's capacity based on the customer's forecast of upcoming needs.

Capital Expenditure Amount of money spent for a major item in a factory, such as a piece of machinery, a building, a piece of furniture, and so on.

Cause and Effect Diagram *See Ishikawa Diagram.*

Cause and Effect Diagram with Cards (CEDAC) An advanced form of a cause and effect diagram developed by Ryuji Fukuda that allows on-line workers to change the causes by a system of cards.

CD-ROM Laser memory device for computers known as a compact disk read-only-media and stores large amounts of data.

Central Processing Unit (CPU) The main unit inside of a computer where the decision making and the processing of information occur.

Chapter 7 A serious bankruptcy situation that places a firm under court jurisdiction with the possibility of liquidation.

Chapter 11 A less serious bankruptcy situation that allows a firm in financial trouble to retain some control and also provides some relief of the payment schedules to the suppliers.

CITE Program A comprehensive quality program that emphasizes employee participation and PCA responses to out-of-control conditions. CITE is an acronym that stands for *Continuous Improvement Towards Excellence*, a program invented at Sigma Circuits.

Cleanliness Being free from dirt or contamination. *See Purity.*

Common Cause Refers to the natural variation of a process.

Component   Part of the whole product that is either manufactured or supplied prior to the shipment of the final product.

Components Search   Design of experiment technique where key components of a good assembly are switched with those of a bad assembly with the differences monitored closely after each switch. The components that produce pronounced effects upon switching are identified as key problem areas.

Composite Material   Substance comprised of two or more separate materials that are bonded together to achieve a combination of properties of the individual materials, particularly strength.

Computer Aided Design (CAD)   Technique of drawing on a computer screen and utilizing a computer's math skills for more rapid design as compared to conventional drafting table mechanisms.

Computer Aided Engineering (CAE)   Performing mathematical and engineering tasks with the aid of a computer to decrease total engineering time.

Computer Aided Manufacturing (CAM)   All aspects of preparing a product to be manufactured with the use of a computer.

Computer Integrated Manufacturing (CIM)   Factory technique of tying in various machines and processes with one central computer control and data acquisition system that allows frequent computer communication throughout the factory.

Computer Numerical Control (CNC)   Older term that refers to computerized control of machines.

Concurrent Engineering   A faster type of design engineering where the design, the manufacturing process, the materials, and the manufacturing tools are worked on at the same time to reduce manufacturing cycles and increase the likelihood of trouble-free manufacturing. *Synonym: Simultaneous Engineering.*

Conditioning Manufacturing   *See Process Manufacturing.*

Consumer Goods   Manufactured items such as bicycles or cameras sold directly to the consumer in retail stores and intended for personal or household use. *See Producer Goods.*

Consumer Price Index (CPI)   A monitoring index by the U. S. Bureau of Labor Statistics that measures the U. S. consumer prices on a monthly basis. *See Cost of Living Adjustment.*

Continuous Improvement Process (CIP)   Factory approach to improving customer service and the product by incremental approaches that focus on the root sources of defects and variability. Even small improvements are considered important and add up to substantial strides in the business both in quality and profitability.

Continuous Flow Processing   Type of process manufacturing where each individual part or component is processed individually in a serial fashion. *See Batch Processing.*

Continuous Measurable Improvement (CMI)   System of measuring the gradual improvement of a system or a product. *See Kaizen.*

Control Chart   Graphic representation of the variability of a manufacturing process element with specification or control limits illustrated as lines on the chart. Points on the chart illustrate data of the variable being measured, such as tire pressure in psi. Utilized to observe and reveal underlying patterns in the manufacturing process so that the variability in the process can be reduced through actions on undesirable patterns.

Control Limits   Lines on a control chart defining the process capability limits. The location of the lines is calculated by the actual process data.

Control Plan   Document that defines all the primary product and process characteristics that contribute to variability. This document also lists the measurement method and a reaction plan in case the variability exceeds the normal process variation. *See Reaction Plan.*

Conveyor   A mechanical moving belt for transporting material from one location to another.

Cosmetics   Appearance of the product, especially as it relates to the final finish, paint, color, and/or workmanship.

Cost of Goods Sold   Accounting term for the total of the direct material, direct labor, and the overhead, and excluding general sales and administrative costs.

Cost of Living Adjustment (COLA)   Common term for wage adjustments based on movements in the general economy and as defined by the Consumer Price Index (CPI). *See CPI.*

Critical Path   Manufacturing sequence that represents the shortest possible pathway for manufacturing the product.

Critical Ratio   The time remaining to manufacture a product divided by the work time remaining. Generally, this concept emphasizes processing the order with the lowest critical ratio first. A ratio below one is past due.

Crosby, Philip B.   A leading American quality expert and consultant whose book *Quality Is Free* was a landmark work on the importance of quality in American manufacturing. He was the first writer to actually state that quality in products improved profits and reduced costs.

Cross Training   The instruction of a particular operator(s) or employee(s) on more than one process or assembly operation. *See Job Rotation.*

Cure   State of a material after it has finished its heat treating cycle.

Current Asset   Type of asset that is likely to be sold or exchanged during a one-year period such as cash, inventory, or accounts receivable. *See Fixed Asset.*

Customer   Person or organization that receives the products or services provided by a company or firm.

Customer Requirements Planning   Further extension of MRP system, which allows the customer full access to reserved capacity and a supplier base at the home factory. In this system the customer considers the factory as one of its own work centers and schedules it accordingly.

Customer Satisfaction General term for successful fulfillment of the customer's expectations.

Customer/Supplier Relationship Team concept involving a dynamic working relationship between a supplier and a customer that emphasizes two-way communication and two-way service.

Cycle Time Average time a particular product requires for assembly during the manufacturing process, not including queue time consumed prior to starting the operation. See *Queue Time*.

Data Measurements of a product or process that can be expressed as a numerical value (variable) or go/no go quantities (attribute).

Database Collection of facts and figures of a manufacturing operation residing in a computer's memory that can be sorted and/or collected in a variety of ways depending on the user of the system.

Date Code Type of lot number that is labeled by the exact manufacturing date and is sometimes used to indicate an expiration date on perishable materials.

Defect An imperfection or shortcoming in the manufactured product.

Defects Per Unit (DPU) The number of defects in a finished product.

Deming's 14 Points of Management W. Edwards Deming's philosophies are exhibited in 14 items for managers to follow.

Deming, William Edwards Encouraged the use of statistical methods to measure quality manufacturing. Also credited as a major contributor to Japan's manufacturing prowess. Began career working for the Census Bureau during WW II with statistical quality control methods, then worked with the Japanese after the war. The Japanese later started the *Deming*

188 Appendix One / Glossary

Prize for those companies who excelled in the use of statistical methods.

Dependent Variable   Process characteristic that reacts or follows the pattern of another process parameter. *See Independent Variable.*

Depreciation   A financial term to describe the wearing out of equipment, machinery, or a plant (fixed assets), and the associated loss in value of these items. Also the financial method of deducting this loss in value on a month-by-month basis.

Design Failure Mode Analysis (DFMA)   Study of all possible potential failures of a product and the possible consequences of each failure. A sophisticated approach for evaluating how much damage Murphy's Law can create in a product. *See Failure Mode and Effects/Analysis.*

Design for Dissassembly (DFD)   Technique of designing products so that after the product life has been expended, the product can be easily taken apart, and the materials can be recycled.

Design for Manufacturing and Assembly (DFMA)   Design discipline that requires designs to be compatible with standard manufacturing processes, thereby achieving high yields and quality product during the manufacturing cycle.

Design of Experiment (DOE)   A formal method of troubleshooting and problem solving that utilizes a controlled experiment to discover the important variables in a product or a design. A change in these critical variables may significantly reduce variation.

Detection   Method of manufacturing product by inspecting all products and removing only the defective products prior to shipment. This method does not promote the more important method of prevention.

Diagnostic Journey   J. M. Juran's approach to analyzing and troubleshooting manufacturing problems. *See Juran., J. M.*

Digital Computer   Most common type of computer utilized. Operates with computer code that is comprised of individual and discrete bits or bytes of data.

Digital Data   Customer information transmitted to a manufacturing plant in the form of computer code and usually generated by CAM or CAD or CAE methods.

Direct Labor   The percentage of labor costs in a product as compared to the product's selling price. Labor costs include only those of the individuals actually assemblying the product and not including the general management, maintenance, inspection, or sales individuals.

Direct Material   The percentage of raw materials in a product as compared to the product's selling price.

Discrepancy   Problem occurring in any aspect of the manufactured product.

Distribution Requirements Planning (DRP)   System for organizing the production of material for distribution channels based on customer forecasts of needed product.

Distributor   Person or organization that buys the product directly from the manufacturer, then later resells the product to another organization for a profit.

Dock to Stock   Concept of supplying product of such high quality that inspection by the customer is not necessary and only adds to the costs of the product. The customer is totally confident in the manufacturer for the quality of the product.

Downtime   Length of time or percentage of time that a particular machine or process is not operating for any reason.

Drawing   Graphic picture of the component or product in a variety of different views and usually including measurements of the key features of the product.

Durable Goods   A manufactured product that lasts for a long time, such as an automobile. *See Nondurable Goods.*

Early Supplier Involvement (ESI)   Program of supplier management that emphasizes the supplier be an actual part of the business and be part of the decision making process of problem solving and continuous improvement.

Economic Order/Build Quantity (EOQ/EBQ)   A method of determining the optimum lot size by analyzing output vs. set-up time and selecting the lot size that provides the lowest total cost of manufacturing. Modern theory states that the ultimate factory would have an EOQ or EBQ equal to one due to advances in set-up time and continuous processing.

Electronic Data Interface/Interchange (EDI)   A computer link between a company and either a supplier or a customer that allows transfer of electronic information, such as purchase orders, quality reports, and so on.

Electronic Design Automation (EDA)   Process of using computers for the design of more sophisticated semiconductors where several semiconductor functions are preprogrammed and can be selected simply to generate more difficult semiconductors, such as gate arrays, and so on.

Electronic Mail   A message service for individuals involved in a computer network system that requires no paper and relies on screen messages.

Electronic Math Data   An all math approach to describing all aspects of a part number including its dimensions

Emergency Response Training (ERT)   Training program for evacuation, first aid, and/or containment of an emergency by all selected individuals in a factory. ERT usually includes an emergency organization with positions such as fire chief, and so on as part of well-defined plan of action.

Employee Involvement (EI)   Concept of including all the employees in some aspect of the business or manufacturing process.

Employee Strategic Participation (ESP)   Program of having employees participate more fully in the long-term planning of the business.

Empowerment Trend in organizations that distributes the decision making process throughout the organization and encourages reasonable and informed risk taking.

End User The customer of the product manufactured.

Engineering Change Level (ECL) Designation of the latest revision of a part and usually indicated by a number or letter that follows the part number.

Engineering Change Notice (ECN) *See ECO.*

Engineering Change Order (ECO) A formal instruction that changes the design or manufacturing steps and is issued by the engineering department.

Environmental Stress Screening (ESS) Type of reliability testing provides an assessment of how a product will perform in the future under a variety of environmental conditions including vibration, heat, humidity, and so on.

Equity Value of a manufacturing company assuming all outstanding financial obligations are met.

Ergonomics The study of the interaction of the human body muscular system and a machine or piece of equipment. The purpose of the study is to develop improved machine interfaces that result in less effort and fatigue for the individual operating the machine.

Experimental Design *See Design of Experiment.*

Facilitator Key individual in a team meeting who is responsible for promoting team interaction and communication. This individual is usually professionally trained in communication skills, human resources, and psychology.

Facilities Buildings, land, and structures that contain and provide utilities to the manufacturing operation and to its equipment.

Facilities Engineer   Individual whose responsibility includes the design of a building for manufacturing, its utilities, and the internal layout of the plant for specific operations.

Facsimile (FAX)   Machine for transmitting documents through standard telephone lines. Also refers to the paper copies of the documents produced at the receiving end.

Factory Loading   Comparison of the actual capacity of the factory and its present orders on the floor, usually expressed as a percentage utilization or a percentage loading.

Failure Mode and Effects/Analysis (FEMA)   Study of all possible potential failures of a manufacturing process and the possible consequences on the end product. *See Design Failure Mode Analysis.*

Federal Trade Commission (FTC)   U. S. government organization dedicated to assuring fair competition.

Feet Per Minute (FPM)   Measurement of the speed of an assembly line or conveyor as expressed in how many feet of product or conveyor would pass a single point within one minute.

Feedback Loop   Manufacturing information that is transferred backwards to the previous work station and is used to signal production prompts or signal the need for improvements in the process.

Field   Refers to all locations outside of the factory that are potential destinations for the product being utilized.

Field Inventory   Term for unsold product that has been prebuilt and located outside the factory and close to the customer's locations. This product is expected to be purchased in a short period of time by the customer, and can provide quick deliveries for the customer.

Filtration   Process of removing dirt and contamination from air or liquids by passing the air or liquid through a filter media such as fibers or charcoal.

Finished Goods Completed product that has not shipped from the facility yet.

Firmware Software such as a semiconductor chip that has been permanently placed inside an electronic component.

First Article Beginning piece of a new process or product that becomes the test piece for quality and performance checks prior to more manufacturing.

First In First Out (FIFO) Accounting term that values the inventory in line with the oldest purchases made. Used in manufacturing to indicate the use of the oldest inventory first. *See LIFO.*

Fiscal Year Accounting term for the 12-month period in which the profit and loss is determined and reported and which does not necessarily correspond to January 1 through December 31. Companies often have fiscal years that do not correspond with calendar years.

Fishbone Diagram A cause and effect tool developed by Dr. Kaoru Ishikawa, an authority on quality control in Japan. The fishbone diagram is a graphic representation of all possible causes of a particular problem. Each spur on the skeleton indicates a possible cause. The head of the skeleton represents the problem.

Fixed Asset A type of asset such as plant, machinery, and equipment that lasts a long time and is not sold or consumed during normal business transactions.

Fixed Costs Expenditures or costs that do not change with the volume of the units running in the factory. The building's lease rate would be an example of a fixed cost. *See Variable Costs.*

Fixture Tool for holding materials or components while a manufacturing operation takes place.

Flexible Manufacturing System (FMS) New factory concept of providing machines, tools, systems, and technologies for the manufacturing facility that requires little to no set-up time between P/Ns and a fast product development cycle, thereby allowing quick changeovers between different products and more adaptability of the factory to the marketplace.

Flexible Manufacturing Technology (FMT) *See Flexible Manufacturing Systems.*

Floppy Disks Computer memory device that utilizes flexible magnetic discs for the storage of digital information. Floppy disks are utilized primarily on smaller personal computers due to their low cost and flexibility.

Focused Factory A factory that manufactures a specific product that has been optimized to achieve very low costs due to this focus on a particular product through specific tooling and layouts.

Ford, Henry A leading manufacturer of automobiles in the early 1900s who revolutionized manufacturing with the use of the conveyorized assembly line.

Forecast A prediction of future requirements for product, services, materials, or components; usually from the customer or the sales department.

Forklift Truck Moving vehicle that has two metal prongs for lifting pallets of material to and from shelves and other trucks.

Fortune 500 Refers to the largest 500 manufacturing corporations in the United States and is published by *Fortune* magazine.

Four Tigers Refers to the four Asian countries of Taiwan, Hong Kong, Korea, and Singapore.

Frontline Manufacturing Continually improves product value by emphasizing and promoting individual contributions to the design, the material, and the process. Promotes innovation, value analysis, waste reduction, lean design, robust process, making it right the first time, and strong supplier/customer relationships through continuous education and participation of all workers.

Full Factorials Exhaustive design of experiment technique that evaluates all possible combinations of variables to determine their interactions.

Fuzzy Logic   An internal computer decision making process that simulates how a human deals with a problem that is not clearly black or white but rather is *grey*. Fuzzy logic semiconductors are now being utilized in consumer products to enhance the *in-between* decisions of running a household.

FYI   For Your Information

Gage   An instrument device that measures process or product variables such as pressure, temperature, length, width, and so on, and provides a numerical value of the variable that can be recorded.

Gage Accuracy   Comparison of the average measurements on a gage to an extremely precise reference gage (such as an NBS standard).

Gage Linearity   A term to describe the uniformity of the gage readings over the entire range of measurements.

Gage Repeatability   Variability in measurements on a gage attributed to a single operator's technique.

Gage Reproducibility   Variability in the average measurements of a gage attributed to differences in operator techniques.

Gage Stability   Variability of gage over time.

GANTT Chart   A graphic representation of project milestones and events containing symbols for project activity start and completion.

Garbage In/Garbage Out (GIGO)   Common term for the problem of entering bad data into a computer database, thereby causing the generation of false reports.

Genealogy   An accounting of the sources of inventory that go into a product including the entire history of how the product came to be manufactured.

General Ledger   An accounting document that lists all credits and debits of a specific business.

General Physics Corporation   Organization aligned with General Motors that helps develop the *Targets for Excellence*

Program of continuous improvement for GM. *See Targets for Excellence.*

Generally Accepted Accounting Procedures. (GAAP) Broad-based documentation of accounting rules and procedures that are updated regularly by the Financial Accounting Standards Board.

Geometric Dimensioning and Tolerancing (GD&T) Newer standard of defining measurements on mechanical or engineering drawings (American National Standard).

Gigabyte Large storage capacity in a computer memory representing one billion units of information.

Global Integrated Companies (GIC) Organizations that have taken full advantage of world class manufacturing and have sites around the world in various locations.

Goal A well-defined target or accomplishment that is desired.

Golf Ball System Method of controlling and limiting inventories by using color-coded golf balls to signal work-in-process or components to move to the next work station or to replenish supplies just utilized. The signal is based on completion of a key assembly or process, and this system is used in the automotive industry in Japan. *See Broadcast System.*

Go/No Go Gages Instruments used to determine if a part is within specification limits by two consecutive attribute tests, one for the high side of the spec limit and one for the low side of the spec limit.

Gross Domestic Product (GDP) Financial index of the value of manufactured goods and services in the U.S.

Hard Disks Computer memory device that utilizes a rigid magnetic disc for the storage of digital information. Hard disks are the primary storage mechanism for the memory of larger computers. *See Floppy Disks.*

Hardware Term for the various components of a computer system such as the disk drives, the keyboard, the monitor, and the main computer. It does not include the software.

Heijunka Japanese term for production leveling or smoothing by balancing the marketing effort to fill a factory. During lean times the market effort would be stronger than during normal times in order to level load the factory.

High Efficiency Particle Filter (HEPA) For removing dirt and dust from the air prior to entry to a clean room.

High Reliability Product that can be used by the customer for long periods of time or with a lot of abuse without failing.

Histogram A bar chart illustrating frequency distribution of a series of measurements. The height of the bars on the histogram represents the frequency of occurrence, and the width of the bars represents the particular class that a measurement falls into.

Hold the Gains Concept of retaining the business knowledge, technology, training, and other intellectual information developed and acquired during the growth of a manufacturing firm.

Hoshin Kanri Planning system in Japan that follows W. Edwards Deming's approach, which starts with broad long-term strategic objectives to serve as guidelines. Shorter range objectives are developed by teams throughout the organization.

Incremental Improvement Concept of getting a little better with each day in manufacturing a product or providing a service. *See Continuous Improvement.*

Independent Variable Process item that is critical and causes other items to change directly with it.

Industrial Engineer Individual whose function is to improve productivity in a manufacturing plant by utilizing such skills as plant layout, previous history in manufacturing, new and improved tools and processes, time studies, and so on.

In Process Inspection Newer method of inspecting product in the factory by the operator or work center that performed the work on an inspect-as-you-go basis and in the area where a correction and feedback can both occur quickly. This is opposed to the older method of having a separate QC department inspecting the product at the very end of the process, a point at which no corrections are possible and feedback is very slow.

Input/Output Yield   Numerical measurement of the total percentage of raw materials that eventually is shipped in the end product in relationship to the total raw materials delivered to the factory.

Integrated Circuit (IC)   Electronic component consisting of several individual active and passive silicon circuits and utilized for electronic signal processing.

Integrated Services Digital Network (ISDN)   Computer protocol that allows for voice, video, and data to be transmitted on the same line.

Measuring Instrument   A measuring device for determining variable or attribute data. *See Gage.*

Intellectual Property   Collection of processing knowledge, skills, procedures, patents, manufacturing plans that belongs to a manufacturing company.

Intelligent Manufacturing   Type of manufacturing characterized by quick response to problems, the retention of solutions (*see Memory Retrieval System*), high quality levels, low inventory levels, and high employee participation and education, all of which result in an adaptability to the marketplace.

Inventory   The materials, components, subassemblies that comprise the end product in various stages of the manufacturing cycle. Also, a financial account of these materials in their various stages.

Inventory Turns   The number of times during a year that the total inventory on the manufacturing floor is completely replaced with new inventory.

Invoice   Document that is sent to the customer after the product has been shipped and which represents the billing for payment of the product.

Ishikawa Diagram   *See Fishbone Diagram.*

ISO 9000   International standard for total quality control developed by the International Standards Organization in 1987. Describes management systems for making it right the first time including a variety of disciplines, such as manufacturing, test and measurement, handling and packaging, sales and customer service, and so on.

Jidohka   Concept of developing machines that signal to operators when there is a mechanical problem that requires attention and also stop themselves when the problem is serious.

Job Description   A document listing all the responsibilities, duties, and the education required for a particular job. The description is used extensively in hiring and job evaluations. *See Job Evaluation.*

Job Evaluation   A study of each job in a factory to determine the relative importance and the promotional opportunities in the organization.

Job Rotation   Method of training where an individual worker is allowed to expand his or her knowledge by periodically moving through the factory and performing each individual job for a period of time sufficient to learn the job. *See Cross Training.*

Juran, J. M.   Early quality expert in U.S. manufacturing who wrote one of the first quality manuals that is still often referred to (*Juran's Quality Control Handbook*). Juran is also noted for his early work in Japan and for starting the Juran Institute—a school for quality control.

Just-In-Time (JIT)   Technique of reducing the total inventory in a factory by delivering material to the next operation only when the next operation is prepared to process the material. This is accomplished through accurate scheduling or through signaling systems from the next operation. *Other terms: Stockless Production, Kanban, Synchronous Manufacturing, Pull System.*

Just-In-Time II (JIT II)   Advancement of JIT concept that involves the supplier as an actual extension of the company's purchasing department. Invented by Bose Corporation, the supplier's representative actually places orders with his/her own company and acts like a production control individual instead of the traditional sales role.

Kaizen   Japanese term for daily continuous improvement. Statistical methods based on the teachings of Deming are utilized in this system to reduce variability in the process and the product. *Synonyms: Continuous Improvement Process; Incremental Improvement.*

Kanban   Japanese system developed by Taiichi Ohno at Toyota for limiting the number of manufacturing lots on the floor by assigning a limited number of tokens to be assigned to the lots—one token per lot. More lots cannot be placed on the floor until previous tokens are collected from lots that have completed the manufacturing operation. Also movement from one manufacturing station to the next is limited by the issue of a token so that material is *pulled* not *pushed* through the system, thereby reducing work-in-process (WIP).

Keihakutansho   Japanese term for the design goals of lighter, thinner, shorter, and smaller.

Keirutsu   Group of companies that work together and support each other. This concept originally started in Japan with large conglomerates.

Key Control Characteristic (KCC)   An important process measurement that defines the variability of the process.

Key Product Characteristic (KPC)   A product measurement of the variability of important customer design and, product safety requirements, or regulatory standards.

Kit   Collection of parts that go into an assembly.

Kitting   Process of collecting all the parts for assembly.

Last In First Out (LIFO)   Accounting term that values the inventory in line with the most recent purchases. Different companies use either the LIFO or the FIFO system depending on their type of business and their accounting system. *See FIFO.*

Lead Time Amount of time between when an order is placed and when the product is shipped.

Lean Design Term for minimizing the complexity of a product by reducing the length, width, depth, weight, and number of components. Reduces the number of components in a product through consolidation of functions and the use of modular parts. *Other term: Keihakutansho.*

Lean Production Comprehensive approach to manufacturing that minimizes wasted motion, wasted materials, and excessive inventory; and emphasizes flexibility by quick tool changes and small lot sizes. Sometimes described as a method that is superior to mass production and requires half the space, half the time to manufacture product, and half the time to develop a new product. Term coined by John Krafcik and describes many aspects of the Toyota Production System. Reference: *The Machine That Changed the World*, James P. Womack, Daniel T. Jones and Daniel Roos, Rawson Associates, NY, 1990.

Liability Financial term that indicates the amount of assets in a company that are obligated to be paid out in the future for goods or services.

Liquid Asset An item of value that can be sold or exchanged for cash within a short period of time.

Local Area Network (LAN) Computer hardware and software system for sharing information among a number of terminals or personal computers.

Location Diagrams Drawing of a product or component with the locations of defects or problems highlighted on the drawing. Utilized to identify patterns of defects and possible causes.

Lock and Key Design Method of design that assures critical components; can only be assembled in one direction, thereby preventing incorrect assembly. *Other term: Poka-Yoke.*

Lock Out/Tag Out   Safety term describing a warning and security system (usually a padlock or a warning tag) to prevent any operation of machinery or electrical equipment by anyone other than the individual authorized to operate the equipment. Lock out/tag out may occur, for example, when a maintenance mechanic is repairing a piece of equipment and activation of the equipment during the maintenance procedure would result in serious injury.

Logo   A trademark or distinguishing symbol that a company uses for immediate recognition of its products or services.

Long-Term Process Capability   Defines the range of the *drift* that may occur in a process over time due to process fluctuations, different operators, different materials, or tool changes.

Lot Number   Method of identifying manufacturing inventory that links a numerical code with the entire history of the inventory in such a manner that all aspects of the operation can be traced. *Synonyms: Date Code; Serial Number.*

Lower Control Limit (LCL)   Bottom value that a process indicator can go to before a correction is made in the process, or the process is shut down for repair.

Lower Specification Limit (LSL)   Bottom value that a process indicator can go to before going out of specification.

Machine Capacity   Maximum output of a machine and/or process assuming there are unlimited operators to run the machine and an unlimited supply of materials or components.

Machine Vision   Technique of using video cameras linked with computers to control and monitor machines during their operation.

Maintenance   Routine repair and lubrication of machines to keep them in proper running order.

Make-To-Order   A factory order method that is based solely on actual customer orders and does not allow orders for field inventory replacement. Each unit on the factory floor will have a corresponding customer order associated with it. *See Make-To-Stock.*

Make-To-Stock A factory order method based on maintaining field inventory and with little to no actual customer orders in the system. *See Make-To-Order.*

Make vs. Buy Decision making process that compares the costs of buying a product to making a product.

Malcolm Baldrige National Quality Award A quality award given to American companies for outstanding quality performance. The award was initiated as an Act of U.S. Congress and is given by the Department of Commerce in dedication to the late Malcolm Baldrige, Secretary of Commerce.

Manufacturing The process of making a valuable product from a design and one or more raw materials or components.

Manufacturing Cell *See Work Cell.*

Manufacturing Engineer Individual who has been trained in all aspects of improving the manufacturing process through better organization, tooling, materials, and so on.

Manufacturing Planning and Control (MPC) Factory system for computerization of the entire planning process for the factory including capacities, equipment, production control, and so on.

Manufacturing Resource Planning (MRPII) Extension of MRP I system that includes the items of personnel staffing and machine capacity to the planning system and is based both on actual orders and a forecast of the customer's requirements.

Market Listing of all the potential customers possible for a product or a grouping of products. This listing usually includes the total amount of dollars spent by all customers in a particular area and the quantity of products purchased.

Market Share Percentage of the total market that a manufacturer controls or supplies into.

Marketing   Process of analyzing potential customers for a product in terms of their needs and includes the continuous planning for growing the customer base and the total customer usage.

Mass Yield   *See Input/Output Yield.*

Master Schedule   Priority plan for the sequence of manufacturing products that includes past dues.

Material Handling Equipment   Machines such as forklifts, conveyors, and robots used to transfer product or material from one manufacturing operation to another.

Material Requirements Planning (MRP I)   System for ordering or buying raw materials or components based on actual orders, and a forecast of the customer's requirements. *See Min/ Max.*

Material Review Board (MRB)   Group of individuals who get together and review the product that has been rejected at any step of the manufacturing process and decide whether the customer would be able to use the product or whether the product should be discarded. This group usually works with those items in manufacturing that are very costly either in dollars or in schedule.

Material Utilization   Percentage raw material used in the final product assuming zero scrap for workmanship. This term usually refers to the material that is trimmed off or never used due to material size requirements.

Mean   Centermost number of a group of numbers or readings. *See Average.*

Mean Time Between Assists (MTBA)   Average time between an incident that requires a maintenance person or a skilled operator to adjust a process in order to make it work correctly.

Mean Time Between Failures (MTBF)   Average time it takes a product to fail in the field with the customer.

Mean Time To Repair (MTTR)   Average time to repair a piece of machinery or equipment when it breaks down.

Median   The center or middle value of a series of measurements.

Megabyte   Large storage capacity in a computer memory representing one million units of information.

Memory Retrieval System (MRS)   Computer listing of critical factory parameters that have been stored and are periodically displayed for audit by individuals within the factory. Includes both the standards and specifications for the factory.

Metric System   Decimal system of weights and measures based on the meter.

Metrology   Science of measurement.

Micrometer   Precision instrument for making very small measurements of distance or size.

Min/Max   Inventory system for ordering raw materials or components based on a fixed amount of stock that varies from a low amount to a high amount. When the inventory falls below a lower amount that is predetermined, more inventory is ordered. In this system the inventory is not related to direct or anticipated customer orders. See MRP I.

Mission Statement   A concise one-page description of a company's purpose and quality philosophy. It is distributed widely to customers, suppliers, and employees.

Modem   An electronic device linking computers via a normal telephone line.

Modular Design   Technique of reducing part counts in a product by encouraging interchangeability of components.

Moving Range Chart   A less common manufacturing chart that is utilized when only one measurement is taken at a time (for example a temperature of an oven). The single measurement is plotted horizontally, and the range is plotted directly beneath. Since only one measurement is taken at a time, the range is calculated as the difference between the current reading and the previous reading.

Muda   Term for the wasted time, materials, and effort in manufacturing and coined by Taiichi Ohno at the Toyota Corporation.

Multiple Environment Overstress Tests (MEOST) A series of accelerated reliability tests for products that simulates long-term life cycles. *See Accelerated Life Testing.*

Multi-Vari Charts Design of experiment technique where measurements are charted in a manner that indicates the variation as positional, cyclical, or temporal, thereby facilitating problem solving.

Murphy's Law Common term that refers to the adage: What can go wrong, will! It describes the degenerative process that often accompanies manufacturing processes that lack adequate controls and planning.

National Institute of Standards and Technology (NIST) Federal agency that provides measurement standards for a variety of manufacturing instruments and tools.

Natural Variation Natural tendency for similar products, materials, and processes to be slightly different within a certain range of values. *See Uncontrolled Variation.*

Net Earnings *See Net Profit.*

Net Income *See Net Profit.*

Net Profit Financial term describing income left over after all obligations (total costs) have been deducted.

Net Sales Financial term describing the sales figure in dollars minus any returns, discounts, or transportation charges.

Next Operation As Customer (NOAC) Nontraditional method of manufacturing that defines the receiving work station as the customer, and the sending work station as the supplier, thereby emphasizing the importance of each individual and his or her respective work cell.

Nikkei Prize Japanese Prize for excellence in manufacturing.

Nondurable Goods A manufactured product such as a food product that lasts for a short period of time and is consumed. *See Durable Goods.*

Normal Curve   *See Normal Distribution.*

Normal Distribution   A graphic or pictorial presentation of the normal variability of a manufacturing process. This distribution has the shape of a bell, and its spread and its center location are calculated with statistical analysis of the process measurements. *Other Terms: Bell Curve, Normal Curve, Standard Distribution.*

NP Chart   A special case of a P chart where the total number of product units found defective in an inspection lot are plotted instead (as opposed to the percentage found defective). This chart requires all lots to be of the same quantity. *See P Chart.*

Ohno, Taiichi   Toyota founder of the Just-In-Time production system. *See JIT.*

On-The-Job Training (OJT)   Most common method of training where a new employee spends time with an experienced employee who demonstrates the proper methods and procedures for the job.

Organizational Chart   Graphic drawing of the key positions in a company, usually with the positions in order of importance from top to bottom.

Other Than Japan (OTJ)   Countries in the Far East (such as Korea, Taiwan), not including Japan.

Out of Tolerance (OT)   Condition when a measurement on the product or the manufacturing process does not meet the specification.

Overage   Parts that have been built without any purchase orders due to production control or manufacturing errors. *See Finished Goods.*

Overhead   Percentage of costs of the final product that refers to the benefits of the direct labor force and the extra costs for inspection of the product.

P Chart   A common control chart for attribute data (go / no go) that plots the percentage of product units found to be defective over time. Space is available on the chart for control limits and the quantities inspected and rejected.

Packaging   Materials and boxes that are placed around the product prior to shipment to protect the product from damage during transit.

Packing Slip   Documentation sent with the product during transit indicating the contents within the packaging.

Paired Comparisons   Design of experiment technique where one pair (defined as one good and one bad component) is analyzed for its differences. Subsequently, several other pairs are analyzed for their differences, and the results analyzed to determine the variables responsible for the differences.

Pallet Jack   Manually operated hydraulic machine for the movement of pallets by the same two-tong system used with forklifts.

Pallets   Platforms that are usually wood and hold product for easy transfer by pallet jacks or forklifts.

Paradigm   A pattern or consistency to a series of operations.

Pareto Charts   Special form of histogram chart that illustrates the order of importance of process scrap. The first bar on the graph represents the factor responsible for the highest defect rate, then each remaining bar represents the remaining sources of variation in descending order.

Pareto Analysis   A study of a process to determine the sources of variation and/or scrap in order of the highest to lowest.

Parts Per Billion (PPB)   Term for one defective part in 1,000,000,000 or 0.0000001%.

Parts Per Million (PPM)   Term for one defective part in 1,000,000 or 0.0001%.

Past Due   Parts or components that have not shipped on the day or the hour the customer requested the shipment.

Periodic maintenance (PM)   *See Preventive Maintenance.*

Perpetual Inventory System that can account for 100 percent of the parts at any moment in time due to extreme accuracy of individual transactions and does not require a physical inventory count.

Personal Protective Equipment (PPE) Protective clothing or equipment such as gloves, helmets, shoes, and eye or ear protection, designed to prevent contamination of the product or to protect an individual from a safety hazard.

Physical Inventory Term for doing an in-person count of the inventory at the end of an accounting period and not relying on accounting records. *See Perpetual Inventory.*

Pie Chart Type of chart that illustrates proportions of the whole by dividing a circle into *pieces of the pie. See Charting Appendix.*

Plan, Do, Study, Action (PDSA) Deming methodology of planning an operation correctly, carrying out the operation, studying the operation for its performance, and correcting the operation with an action if required.

Poka-Yoke Method of design that makes certain aspects of the product foolproof for the customer or for the operator in a manufacturing facility. Method emphasized by Shigeo Shingo, a consultant to Japanese Manufacturing firms.

Precontrol An early warning system for shutting down the manufacturing process that does not require an understanding of mathematics, but rather depends on an operator recording data points in one of three specification zones—the green zone for acceptable; the yellow zone for caution; and the red zone for unacceptable. When two or more yellow values are recorded, or when a red value is recorded, the process is to be shut down and corrected. Developed by Frank Satterthwaite, a statistician. *See Appendix.*

Predictive Maintenance A forward thinking, proactive maintenance program that attempts to maintain equipment before an anticipated breakdown or unscheduled interruption.

Prevention   Method of manufacturing that promotes *making it right the first time*. It involves extensive planning and achieving process capability to produce 100 percent acceptable product.

Preventive Maintenance (PM)   Work performed on machines on a regular time schedule to ensure that they always run correctly. For example, PM performs lubrication and replacement of key parts on a machine prior to failure. *See Predictive Maintenance.*

Pride of Workmanship   The new American work ethic that emphasizes high quality products, intelligent manufacturing, and a spirit of involvement by all workers in the manufacturing process and product.

Proactive   Style of running a factory where personnel react to a problem situation or an opportunity before it occurs.

Probability   Mathematical definition of the likelihood of a given event to occur. For example, the probability of a *tails* flip of a coin is 50 percent since the event is one outcome out of a total of two possible outcomes.

Procedure   Well-defined and documented sequence of steps to be followed in manufacturing in a defined work cell. *See Standard Routing.*

Process Average   The arithmetic center of a grouping of process measurements calculated by adding the values and dividing by the number of measurements taken. *Other Terms: X Bar, Arithmetic Mean.*

Process Capability   The defined range of normal operation of a particular process that is within control, has a normal variation, and has no abnormal special or assignable causes. Process capability can be defined as a mathematical relationship between the spec limits and the range of variability of the process with either the term Cp or Cpk.

Process Capability Index (CPK)   See Capability Performance Index.

Process Capability Study A statistical study of a process to determine the natural variation caused by all common causes and the relationship of the variation to the specification. This study normally uses the terms of Cp or Cpk for defining the process capability.

Process Control Term for the maintenance and adjustment of a manufacturing process through the use of feedback loops, statistical process control (SPC), and standard operating procedures (SOPs).

Process Control Action (PCA) A documented adjustment made in the manufacturing process that reduces variability and is prompted by an action signal.

Process Manufacturing Type of manufacturing that produces goods by transformation of materials either mechanically, thermally, chemically, or by some other process. For example, an aluminum manufacturer will change aluminum ore to aluminum products such as aluminum foil by a series of manufacturing steps.

Producer Goods Manufactured items sold to other manufacturers for inclusion into consumer goods. *See Consumer Goods.*

Producibility Ability of a product design to be manufactured.

Productivity Amount of acceptable products manufactured per the amount of employees in the manufacturing operation. Used as a measure for the continuous improvement of an operation.

Profit Financial term referring to the income from doing business after all costs are removed. Profit is usually expressed as a value with the term *after taxes* or *before taxes* to describe the point in time the profit is disclosed.

Programmable Logic Controller (PLC) Mini-sized computer that is used for low-cost control of processes and instruments.

Proportional Integral Derivative (PID) Type of specialized PLC used in sophisticated process control equipment.

Pull-In   Rescheduling of product so that the product ships sooner than originally committed.

Pull-Out   Rescheduling of product that results in a later ship date.

Pull System   Operators in a manufacturing area signal to the previous operation to *pull* material or components to their station when they are out of work, thereby eliminating the staging or queuing of material. *See Just-In-Time.*

Purchase Order   Written document that contains an accurate description of the product being ordered, the quantity, and the price, given to the supplier as a formal request for product.

Purity   Quality of being free from dust or dirt.

Q1, Q101   Ford Motor Company programs for supplier evaluation and certification. The Q101 is the initial qualification program that requires a substantial amount of employee involvement and statistical process control in order to achieve the minimum score. The higher Q1 rating requires sustained performance over an evaluation period.

Quality   A measure or degree of excellence.

Quality Assurance (QA)   Grouping of individuals giving direction and leadership in obtaining excellence. Also the process of producing quality product through direction and leadership.

Quality Circles   Teams that include all workers from all levels including but not limited to supervisors, engineers, and managers that meet regularly to discuss ideas on improving the quality of the manufacturing process.

Quality Control (QC)   Grouping of individuals responsible for accepting or rejecting parts and product with varying degrees of quality. Also the process of ensuring product quality through inspection.

Quality Function Deployment (QFD)   Quality system that is focused on translating the end customer requirements into measurements for each step of the manufacturing process.

Queue Time   The length of time the work-in-process remains idle before an operation begins. Components within the waiting line are referred as *in queue.*

Quick Turn Around (QTA)   System for supplying product to the customer at the fastest possible rate.

Quincunx   A pin-ball type instrument used to demonstrate variation by the dropping of small beads into vertical channels through a series of pins. Invented by Galton, a mathematician.

R & D   Term for the premanufacturing cycle that develops new products. Also known as Research and Development.

R & R   A repeatability and reproducibility study of the variation of measurements on a part or on a process. *See Gage Repeatability and Gage Reproducibility.*

Range   Numerical distance from the highest value to the lowest value of a measurable item of a manufacturing process. For example, the temperature of an oven may cycle between 85 degrees and 100 degrees during the day and the range for the process would be 15 degrees.

Raw Data   Numerical values taken directly from an instrument, gage, report, or an observing individual and prior to any mathematical or statistical function.

Raw Material   A material or component that is staged prior to the first manufacturing operation.

Reaction Plan   Formal document that lists the actions to be taken for each possible type of failure in factory.

Reactive   Style of running a factory where personnel respond to problems after they occur. *See Proactive.*

Real Time   Measurements that are taken and then acted upon simultaneously without delays.

Receiver   Written document that records when product, equipment, or material arrives at a factory for assembly or for use in the factory.

Red Flag   An indication of a potential problem that should be examined before a major problem disrupts the process or product. *See Warusa-Kagen.*

Reliability   Measure of how long a product will last in the field without failing or requiring repair, and a general measure of quality over a period of time.

Resistance Temperature Device (RTD)   A type of electronic thermometer.

Retained Earnings   Profits that are accumulated from year to year.

Return on Equity (ROE)   Net profit expressed as a percentage of the stockholder's equity.

Return on Investment (ROI)   Net profit expressed as a percentage of the total investment of both the stockholders and any banks or funding operations (long-term debt).

Revenue   Dollar amount of the product sold during a fixed period of time.

Robots   Automatic machines or equipment utilized to assemble product in a factory.

Robust Design   A product design that is easily manufactured because of its simplicity and its excellent compatibility with the manufacturing process and the materials used.

Root Cause   Primary factor responsible for unplanned manufacturing events. *See Assignable Cause.*

Run Chart   The simplest of all manufacturing charts for observing graphically a process variable. This chart illustrates the value of the variable in the Y axis, and a straight line connects each new reading horizontally thereby providing a line chart.

Sales Order   Actual verbal or written documentation that describes the customer's order for product and initiates the production cycle within a factory.

Sales Representative   Individual who receives a commission for selling product for a company but which does not work directly for the company.

Sample   A representative part taken from a larger group or population that is studied and analyzed as a method to better understand the characteristics of the entire population.

Sampling   Statistical term for collecting information about a lot of material by randomly selecting product from a lot of material. Sometimes referred to as random sampling.

Scatter Diagram   Simple type of chart that illustrates one variable on the X-axis and one variable on the Y-axis with each reading shown as a dot. By analyzing the dots, the relationship between the two variables can be visually determined. *See Charting Appendix.*

Scrap   Product that has defects that are unacceptable to the product's customer.

Self-Contained Breathing Apparatus (SCBA)   Portable air supply similar to diving equipment that is utilized in safety situations such as the cleaning of a confined space where the air quality would require the provision of clean air.

Sensor   Instrument that collects numerical information about how a process is running.

Service Industry   Any nonmanufactured or agricultural activity that fulfills a need, such as health care, education, and entertainment.

Service Quality Indicators (SQI)   Measurable indicators of how the customer feels towards the manufactured goods. For example, a SQI would be the results of customer survey describing the quality of the paint finish on a product, and the SQI would be expressed in real numbers such as the percentage of customers pleased with the paint finish.

Set-Up Time   Time required to prepare a work site for a new product to be run.

S,G&A   Sales, general and administrative costs that are separated out and not considered direct manufacturing costs.

Shainin, Dorian   Key American individual responsible for the concept of precontrol and other various statistical techniques such as multi-vari charts and components search.

Shewhart, Dr. Walter   Developer of standard control chart.

Shipper   Paperwork that gives instructions for the proper method of shipping the product to the customer or vendor.

Shop Floor   Refers to the manufacturing area and the proximity to the manufacturing area.

Shop Floor Control   System for computer monitoring and reporting of all the manufacturing area's transactions such as time card control, passage of material from station to station, quality levels at each station, and so on.

Shop Supplies   Type of manufacturing material or component that supports, but does *not* become a part of, the final product. For example, shipping crates are shop supplies.

Shusa   Japanese term for a project team leader in an organization.

Sigma   A numerical measure of the variability in a product or process. Determined by statistical sampling of a population and designated in formulas by the Greek letter sigma (σ). The formula is the square root of the sum of the mean squared average difference between each measurement and the process average. *Other Term: Standard Deviation.*

Simultaneous Engineering (SE)   *See Concurrent Engineering.*

Single Minute Exchange of Dies (SMED)   Method used by Toyota to drastically cut the time required to change tools for making automobile parts from hours down to minutes.

Single Source   *See Sole Source.*

Six Sigma Process   Control program developed by Motorola that sets a goal of a six sigma spread between the target value of the process and the closest specification. The six sigma program allows for a process shift of 1.5 sigma and is equivalent to a Cpk rating of 1.5.

SKU Number   Identification number for products usually sold in retail outlets and linked directly to barcoding identification. *See UPC*

Software Computer instructions that reside in an easily changed format of data storage and require or involve no hardware.

Sole Source A vendor or supplier who has an excellent performance record and is awarded 100 percent of the business on a particular item or service due to this top-notch performance.

Special Cause *See Assignable Cause.*

Specifications Detailed descriptions of dimensions, materials, quantities, temperatures, and so on, used to manufacture product.

Staging Preparation of product in front of a work center.

Standard Measurement point that is fixed and utilized as a reference point for instruments in manufacturing. The standard is usually calibrated by an independent group.

Standard Deviation A numerical measure of the variability in a product or process. Determined by statistical sampling of a population and designated in formulas by the Greek letter sigma.

Standard Operating Procedures (SOPs) Step-by-step instructions for performing a manufacturing process or operation. *See Standards.*

Standard Routing Set of work instructions for a particular product to be manufactured that includes the engineering drawings and specifications.

Standards Rules, procedures, and references that are incorporated into the manufacturing environment and govern the daily operations. Standards usually refer to factory imposed rules as opposed to specifications that refer to customer rules and procedures. *See Specifications.*

Statistical Control Term to describe a process that has all special causes eliminated and the remaining variation from the common causes is within the control limits.

Statistical Operator Control (SOC)   Control and improvement of the processes by the individual operator utilizing statistical methods. In SOC the operator monitors the variability of the process by charting feedback loops and performs process corrective actions when excessive variability occurs. *See Statistical Process Control.*

Statistical Process Control (SPC)   Controlling and improving a process by recording sample readings of the process, analyzing the readings with statistics, and performing process corrective actions when the process starts to drift out of control.

Statistical Quality Control (SQC)   System for defining the acceptance and rejection of parts by defining the variability in the parts utilizing statistical tools.

Statistics   Science of collecting and analyzing data utilizing extensive mathematical tools.

Stockless Production   Method of manufacturing that minimizes the work-in-process by reducing significantly or eliminating the queuing of components or materials. *See Just-In-Time.*

Strategic Partnerships (SP)   Relationships with customers or vendors that are stronger than normal and include a higher degree of trust and communication of information.

Subassembly   A smaller grouping of assembly operations that completes a part that later becomes a component of a larger assembly.

Success-Oriented Characteristics (SOC)   Set of personality traits that enable individuals to fit into work environments successfully and assist promotion into more responsible positions.

Supplier   Person, organization, or vendor that manufactures or sells materials, components, subassemblies, or services to the primary manufacturing firm.

Supplier-Customer Relationship   Team concept involving a dynamic working relationship between a supplier and a customer that emphasizes two-way communication.

Synchronous Manufacturing   *See Just-In-Time.*

Synthetic Manufacturing   A simple type of manufacturing where ingredients are mixed or parts assembled. Such as, assembling a bicycle with various parts such as wheels, chains, and tires to make a finished product. *See Assembly Manufacturing.*

Taguchi, Dr. Genichi   Famous Japanese statistician responsible for substantial development of design of experiments in manufacturing. *See Taguchi Method.*

Taguchi Loss Function   Concept of cost accounting that states a company loses money whenever a process drifts from the target even if the process is still within specification. Developed by Dr. Taguchi.

Taguchi Method   Sophisticated tool for problem solving that uses mathematics to determine the importance of manufacturing variables and their interaction. The Taguchi Method is known to be quick determination of the primary causes of problems and is known to reduce the need for costly experiments.

Targets for Excellence (TFE)   Outstanding program within General Motors Corporation that defines in detail all aspects of continuous improvement at the supplier and manufacturing level. Includes sections on leadership, quality, cost, delivery, and technology, and is the standard for supplier performance evaluation and reporting.

Team Oriented Problem Solving (TOPS)   Modern approach to problem solving performed by a select group of individuals who represent a variety of functions within a factory and who utilize an eight-point problem solving technique that defines the problem, contains it, fixes the problem, then rewards the group.

Technical Competency   Knowing the key elements that are important or critical to a business.

Technology Crossovers   Utilizing a variety of technologies to quickly solve a problem.

Telegraphing   Effect of a change in an early stage of manufacturing that is manifested in subsequent operations.

Theory of Constraints   Idea and practice developed by Eliyahu M. Goldratt state that productivity can be best managed by focusing on the work center that has the least capacity for manufacturing, and scheduling that work center to its fullest utilization. A good reference work on this topic is the *The Goal*, Eliyahu M. Goldratt and Jeff Cox, North River Press, NY, 1984.

Theory X   Management theory that states the necessity of strong authoritarian supervision to produce results.

Theory Y   Another management theory stating that with supportive management, individual employees do not require strong supervision and will be self-motivated and successful in their jobs.

Theory Z   Theory promoting strong participation by all employees in the management of an organization and participation by all employees in the rewards of a successful company. Theory Z was developed by William Ouchi.

Thermoplastic   Plastic used in manufacturing that can be characterized by the fact that it melts when heated to a high enough temperature. *See Thermoset.*

Thermoset   Plastic used in manufacturing that can be characterized by the fact that it does not melt when heated; usually it will degrade if heated too high. Thermosets usually consist of a mixture of two or more materials that cure after mixing. *See Thermoplastic.*

Time Compression   Process of eliminating wasted time in all aspects of manufacturing by performing tasks concurrently.

Time Hook   A software link to the outside world that issues an audit request at a specified interval and requires that the audit be completed and reentered into the computer. If the audit is not completed on time, an alarm is issued from the computer.

Tolerance   Variability in a product or process that is considered acceptable by either the specifications or standards.

Tooling   Mechanical devices that align and hold product during the manufacturing steps

Total Productive Maintenance (TPM)   System for maintaining equipment and facilities to a high degree of reliability by a constant replacement and lubrication of all components prior to failure and throughout the life of the equipment. *See Predictive Maintenance.*

Total Quality Commitment (TQC)   Describes a *culture* of continuous measurable improvement and refers to the solid motivation of all the individuals within an organization to improve the general quality of the business.

Total Quality Management (TQM)   Comprehensive factory management system that measures and evaluates all quality levels of all systems within the organization including *internal customers* such as the finance department, facilities department, and so on.

Totes   Containers for carrying parts or materials in a manufacturing plant.

Toyota Production System   Efficient manufacturing system that emphasizes precision deliveries by suppliers to achieve Just-In-Time status and very quick tooling changes between part numbers. Incorporates the ideas and innovations of a key executive at Toyota, Taiichi Ohno. *See Lean Manufacturing.*

Traveler   Manufacturing instruction list that accompanies the part being manufactured. *See Standard Routing.*

Type I Error   *See Alpha Error.*

Type II Error   *See Beta Error.*

U Chart   A special case of a C chart that tracts the number of defects per unit, and each unit may have a variable sample size. *See C Chart.*

Uncontrolled Variation   Variability traced to unplanned events such as human error, faulty machines, or defective material.

Underwriters Laboratory (UL)   Independent laboratory that certifies, tests, and qualifies materials and/or products for a variety of product safety parameters including flammability and electrical properties.

Uninterrupted Power Supply (UPS)   Battery backup power supply to maintain control equipment during a power failure.

Universal Industrial Code (UIC)   An industrial version of the UPC. *See Universal Product Code.*

Universal Product Code (UPC)   A bar coding system for product tracking in the retail system, administered by the nonprofit Uniform Code Council, Inc.

Upper Control Limit (UCL)   Top value that a process indicator can go to before a correction is made in the process or the process is shut down for repair.

Upper Specification Limit (USL)   Top value that a process indicator can go to before going out of specification.

Uptime   Amount of time that a machine or process is operational as compared to the total time available. Usually expressed as a percentage of the total time.

Value Added Management (VAM)   Management technique that emphasizes the elimination of all waste operations and materials throughout the factory, thereby ensuring that all activities *add value* to the product.

Value Analysis   Study technique utilized to review all aspects of the design and the manufacturing process and evaluate each on the basis of whether the design or the process *adds value* and is not simply a wasted component or a wasted step in the manufacturing process. Originally developed by General Electric in 1947.

Value Engineering   Type of engineering that involves the customer, manufacturer, and supplier in an ongoing design process to achieve maximum *value* of each component and each manufacturing step, thereby reducing the total cost of the final product. This type of engineering seeks to eliminate nonvalue components or steps by designing in functionality.

Variable   Any characteristic of a manufacturing process or of a product that may change. The thickness of a cutting tool is a variable. *See Attribute.*

Variable Costs   Expenditures or costs that change directly with the volume of the units running in the factory. For example, each bicycle sold will require the cost of two wheels and two tires. *See Fixed Costs.*

Variable Data   The numbers derived from quantitative measurements of a variable and reported as numerical values or in statistical format.

Variable Search   Sophisticated technique of determining the most influential variables by first ranking the key variables as to influence, prescribing a best or worst condition to each variable, then running an experiment to verify the influences of each variable.

Variance   *See Variability.*

Vendor   *See Supplier.*

Vendor Quality Rating and Service (VSR/VQR)   Method of assigning numerical values to the different suppliers so that the various suppliers can be evaluated for quality of components and service.

Vertical Integration   A factory that makes most of its raw materials instead of purchasing them from vendors or suppliers.

Warehouse Information Network Standard (WINS)   Electronic data transfer standard for inventories, administered by the nonprofit Uniform Code Council, Inc.

Warusa-Kagen   A term to describe items in manufacturing that are not quite correct but are not yet a problem. These types of items are often targeted first for total quality management strategies. *See Red Flag.*

Waste   Any material, component, time, or expense that is used poorly and has no value to end product.

Waybill   Document that travels with the shipment of the product to its final destination describing the transportation routing and cost.

Work Cell   Location of a sequence of manufacturing operations that can be performed without moving material or components a great distance, usually performed by one operator. A typical product line would have several work cells.

Work-In-Process (WIP)   All material, components, and assemblies between the first and the last step in the manufacturing operation. Usually expressed in either numerical counts or monetary terms.

Work Order   Written or computer-generated request from one department to another for a specific task.

World Class Manufacturing (WCM)   Quality of manufacturing skills and products that is competitive on a world-wide basis so that import and export of goods and products is easily accomplished.

X Bar   *See Process Average.*

X Bar/R Chart   The most commonly used trend chart for manufacturing. This chart plots horizontally the average value (X Bar) of multiple readings and also plots horizontally the range of values. Space is provided on this chart for all the individual readings and the control limits that are usually calculated from the data for the X Bar and the range.

Yield   Amount of acceptable product produced after one or more manufacturing processes. Often yield is described as a percentage of good material to the original material started in the operation.

Zero Defects   Quality concept that places the goal of manufacturing to be a product with no defects.

Appendix Two
Charting Your Course

BASIC CHARTING SKILLS

For charts to be dynamic and revealing, they must be prepared, read, and understood on a daily basis. A good chart is like a map. It basically tells your location in relation to the goal.

The mathematics that accompanies these charts has been left out since it is not as important as the visual interpretation of the data. There are many good reference books on the statistics and calculation of control limits and Cpks, if you would like to learn more about the mathematics utilized to interpret the charts.

You will concentrate on *reading* the trends. Good charts tell you where you've been and where you are now, and they can help you understand where you are going in your manufacturing process. Four rules related to charts are:

- Write on your charts!

- Remember to take the correct sample size.

- Make sure the data are clean and accurate.

- Chart the process and the product.

The most important rule in charting is to write all over them! Charts are made for writing notes and discussing problems. If you are in a shop, and you see nice, clean charts, they are probably not doing much good. Don't be afraid to write down your concerns and actions. Any action that you take should be noted on the chart—even if there isn't an assigned box for it. The charts are for your use.

Secondly, taking the correct sample size that is representative of your process is important when charting. The *sample size* is the number of parts you check as a percentage of the total number of parts. You need to have an adequate amount of data from the parts to do the charts correctly. Some charts require a fixed sample size, some require 100 percent sample, and still others require just enough of a sample to be representative of the whole group. In most cases this sample size number or amount is written on the charts and is provided for you. If you need help, you will need to find a person in your organization who is familiar with statistics.

In addition to the proper sample size, you need clean, crisp data. If the data you use are not accurate and reliable, the charts will be meaningless.

Finally, total control can only be achieved by charting both the manufacturing process *and* the product. Charting the final product is useful, but it is after the fact! The product has already been manufactured by a process that may be out of control. When we studied make it right the first time, we emphasized preparation for manufacturing. The same concept is true in process control. If you chart the characteristics of the manufacturing process and control it closely, you will get a jump start in making good product. By utilizing the charts in the correct manner, the manufacturing process will improve over time. Your customers will love you for it!

The following are some of the key charting methods used in manufacturing. In most cases the charts can be used for either attribute or variable data.

Scatter Diagrams

Scatter diagrams show the relationship between variables. One variable is on the X-axis (left to right), and the other variable is on the Y-axis (up and down) with each reading shown as a single dot. By analyzing the dots, the relationship between the two variables can be visually determined. In the example given below, there is a relationship between the tire pressure and braking distance of a mountain bike. This type of charting is used to determine if relationships exist between variables in a manufacturing process. If there is no relationship, the dots will be random or all over the chart. If there is a relationship, the dots will tend to line up in somewhat of a straight line as illustrated in the following example:

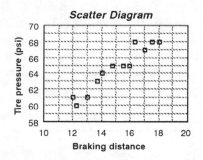

Line Chart

A line chart is a logical extension of a scatter chart. This time a line is drawn between each point so that the relationship can be more closely analyzed. A line chart would be difficult to draw if there were little relationship between the variables; therefore, it is usually used when there is an expectation that there will be a relationship. In the following example, we can conclude that higher tire pressures cause greater braking distance. Based on this chart, Everest Mountain Bikes, Inc. decided to set the tire pressure at 65 lbs. because it wants its bikes to stop easily and be safe in a variety of conditions:

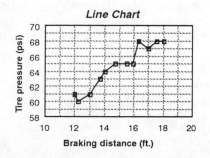

Bar Chart

Probably the most common and easiest chart is a bar chart. In a bar chart, a manufacturing value such as tire pressure is illustrated by a bar instead of a point on a graph. In this manner the measurement is clearly represented. This example illustrates the different tire pressures achieved by each of the individuals who work in the Wheel Assembly department at Everest Mountain Bike, Inc. Each individual working the same machinery achieves slightly different levels of pressure than the others. The variability in the tire pressure can be monitored. Although this chart shows variability, it does not provide many clues regarding how the variability occurred—through natural or assignable causes.

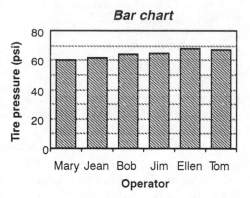

Histogram

This special type of bar chart features bars that represent a series of measurements instead of a single measurement. The height of the bars on the histogram represents the frequency of occurrence and the width of the bars represents the particular class that a measurement falls into. The following is a histogram representing 15 measurements of tire pressure. In this particular chart, the most common tire pressure is 64 lbs.

   It's interesting to note that on the last bar chart of tire pressure, it was not clear as to which pressure was the most common. This type of chart is best when you are trying to discover the most frequently occurring measurements (the central tendency of a group of numbers):

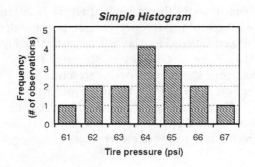

Histogram of Defects

In the previous example, we used a histogram to illustrate a variable of the process, which was the tire pressure being measured in the Wheel Assembly area. Histograms can also be used with attribute data and can visually show how many defects there are in a product. The histogram graphically represents the number of defects in the product. The following histogram shows that the most common defect occurrence is four defects per bike! In other words, five bikes had four defects each. Although this is not good news, it is better to understand the bad news and correct the situation.

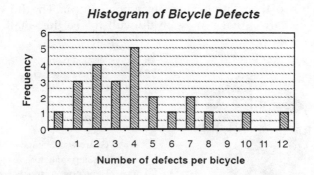

Pareto Charts

A pareto chart is a priority chart. It is a special form of histogram chart that illustrates the order of frequency of defects. The first bar on the graph represents the most frequently occurring type of defect and each remaining bar represents the number of other types of defects in descending order. It is easy to see that the biggest defect at Everest Mountain Bikes, Inc. is missing paint. Since this is the biggest defect, time should be spent correcting it. However, the most frequent defect may not be the most important one. A lower frequency defect may result in a product safety problem and should obviously be attacked first! With pareto charts, you often get a lot of bang for the buck in terms of effort.

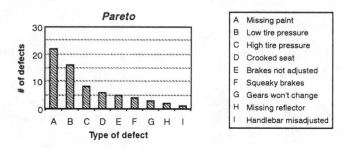

Pie Chart

A pie chart is a totally different approach to analyzing the number of defects in a sample size. It is very effective since it also gives a clear picture of the relationship to the whole sample. The drawback to the pie chart is that it may be difficult to see the smaller pie segments.

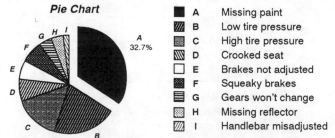

P Chart

The most common chart used for attribute data (go/no go) is the P chart. There are many variations of the P chart such as the NP chart, the C chart, and the U chart. The various types of P charts are used when the manufacturing area wants immediate feedback on the amount of scrap generated as well as the causes of the scrap. Space is available for recording the number and type of defects daily and for tallying the percentage of these defects. Usually a P chart is monitored to determine any upward trends away from the 0 percent line. The 0 percent line is the perfect situation since it means there were no defects for the day. This chart is particularly helpful when a deviation from the norm is noticed by the meandering line—either upward or downward.

Although this type of chart requires a percentage calculation, it is one of the most powerful day-to-day charts for improving a process. It clearly pinpoints the daily problem areas.

P Chart

% Defective	Mon	Tues	Wed	Thurs	Fri
High tire pressure	1	2		5	
Low tire pressure		1		1	
Incorrect bearing seating		1			2
Squeaky wheels		1			1
Gear misalignment	1			1	
Total units defective	2	5	0	7	3
Total units manufactured	100	100	120	100	80
Defect percentage	2	5	0	7	4

Fishbone Diagram

This cause and effect diagram is a graphical representation of a troubleshooting guide that resembles a fish's skeleton. Each spur on the skeleton indicates either a machine condition, the type of manpower, the particular material, the method used, or the environment. The head of the skeleton represents the condition or effect observed.

Also referred to as the Ishikawa diagram, or cause and effect diagram, this is an excellent tool for determining the cause of the variability or failure in a process. Every process should have a fishbone diagram in place so that when problems occur they can be tracked quickly and accurately:

Fishbone Diagram

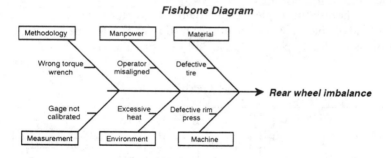

Run Chart

The run chart is the simplest of all manufacturing charts for observing graphically a process variable. The following chart illustrates the value of the tire pressure in the Y axis, and a straight line connects each new reading horizontally. This produces a line chart that gives a picture of how the tire pressure is running throughout the day:

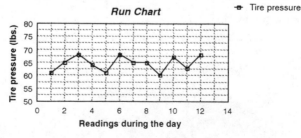

Control Chart

This chart goes one step further and overlays some practical control limits onto the previous run chart, which provides a comparison or reference point to measure against. The control chart is utilized to observe and reveal underlying variations in the manufacturing process. Variability in the process is reduced by reacting to points that approach or move outside the control limits.

There are many variations of control charts utilized in industry today. They include X bar/R charts, X bar/S charts, moving range charts, etc. Fortunately, most of these charts include instructions and examples when provided. The basic control chart remains the same, and the chart type is selected on the application and sample size. Remember, in all the different types of control charts, you are trying to visualize the trends in the data in an effort to reduce variation.

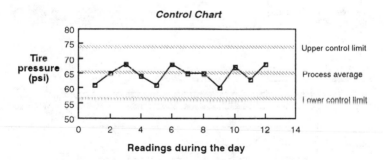

Precontrol Chart

An innovative method that does not require an understanding of mathematics, but allows you to achieve a minimum Cpk of 1.0, is precontrol charting. The chart requires you to plot data points in one of three regions—the green zone for acceptable, the yellow zone for caution, and the red zone for unacceptable. If you can obtain all values within the center green zone, the process is operating at a minimum Cpk of 1.0.

The chart is easily set up. The space between the upper and lower control limit is divided into four equal quadrants. In the case of the Everest Mountain Bike, Inc., these four quadrants represent

5 psi increments from 55 psi up to 75 psi. The green zone is defined as the center two quadrants, and the yellow zone as the outer two quadrants. Finally, the red zone is all areas on the control chart outside the green and yellow zones.

By following some simple precontrol rules, which are listed below, the manufacturing process can be adequately controlled with significant safety limits. When precontrol signals the process to be stopped, this is the same as the previous signals to adjust the process (PCA or a similar adjustment) prior to continuing. The following example illustrates the tire pressure being measured on several tires during the day, and it is based on Everest's specification of 55 to 75 psi being the minimum and maximum values allowed to ship to the customer.

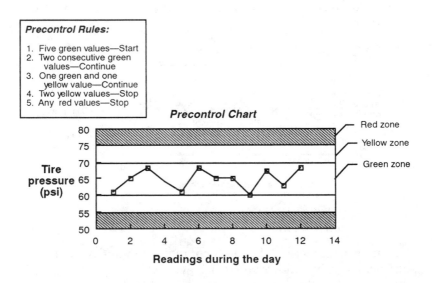

Precontrol Rules:

1. Five green values—Start
2. Two consecutive green values—Continue
3. One green and one yellow value—Continue
4. Two yellow values—Stop
5. Any red values—Stop

Precontrol Chart

Precontrol is easy to understand and use inside a manufacturing facility and requires no calculations. Precontrol was invented by Frank Satterthwaite, a statistician, in conjunction with Rath and Strong, a consulting company for major corporations. This method of control is quite valid statistically, and since there are no calculations involved, this technique is gaining acceptance. For an excellent description of this technique and its use, refer to Keki R. Bhote's publication, *World Class Quality*, Amacom, New York, 1991.

Project Schedule Charting

The GANTT chart is a graphic representation of project milestones and events with symbols for project activity start and completion. This is a simple chart for planning ahead. As you complete tasks, fill in the triangles or darken the lines. This allows you to see your progress over time.

Gantt Chart

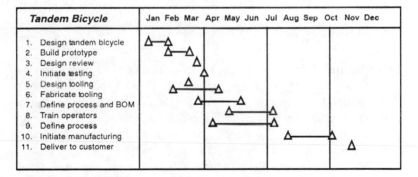

Tandem Bicycle	Jan Feb Mar Apr May Jun Jul Aug Sep Oct Nov Dec
1. Design tandem bicycle	
2. Build prototype	
3. Design review	
4. Initiate testing	
5. Design tooling	
6. Fabricate tooling	
7. Define process and BOM	
8. Train operators	
9. Define process	
10. Initiate manufacturing	
11. Deliver to customer	

Process Flow Charts

Graphic or pictorial representation of the manufacturing steps in a product build includes inspection steps and decision points. As we learned in Chapter 7, this tool is used for analyzing which steps are valuable and necessary in a production process. By eliminating the nonvalue-added steps, time and money can be reduced making the operation more competitive. Process flow charts also provide an excellent training tool for new employees!

Process Flow Chart

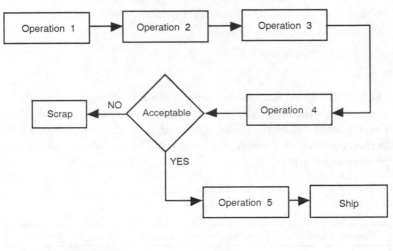

SUMMARY

Don't be afraid to try a variety of graphs and diagrams in your operation. Charting is extremely useful. There is nothing sacred about which technique you use. These techniques aid in feedback loops that make your operation successful. Charting and diagrams provide ways to achieve quick and accurate feedback. Learn as much about charting as you can.

One of the things you will learn is that charting is a universal language. If you know charting, you can walk into virtually any manufacturing facility in the world and read the charts! You will know whether the factory is in control or out of control. You also will know whether it is using dynamic charts, or someone is just fixing the charts for tourists. With charting, you have learned a universal language!

Appendix Three
Frontline Rules of Manufacturing

(1) Maintain pride of workmanship, and manufacture only the highest quality products.

(2) Practice the discipline necessary to manufacture the product correctly the first time

(3) View your coworkers as your customers, and only pass quality products on to them.

(4) Produce better product, and create safe working conditions through attention to detail, and cleanliness of your tools, equipment, and work area.

(5) Communicate problems clearly to other members of the manufacturing team, and learn to shut down the process responsibly.

(6) Keep good product moving by minimizing wasted time, energy, and materials.

(7) Keep the design, process, and material simple.

(8) Continuously improve the manufacturing process and the product.

(9) Continuously reduce the variability in the process and the product.

(10) Center the manufacturing process well within specification limits.

(11) Fix the problem, not the blame.

(12) Continuously improve your personal skills through a positive attitude, team involvement, communication, and education.

Appendix Four

Weights and Measures

Unit	Abb.	Equivalent	Metric
Weight			
ton, short		2000 lb	0.907 metric ton
ton, long		2240 lb	1.016 metric tons
pound	lb	16 oz	0.453 kilogram
ounce	oz	16 dr	28.349 grams
dram	dr	27.343 gr	1.771 grams
grain	gr	0.002285 oz	0.0648 gram
Capacity, Liquid			
gallon	gal	4 qt	3.785 liters
quart	qt	2 pt	0.946 liter
pint	pt	28.875 cu in	0.473 liter
fluidounce	fl oz	8 fl dr	29.573 milliliters
fluidram	fl dr	0.225 cu in	3.696 milliliters
Capacity, Dry			
quart	qt	67.200 cu in	1.101 liters
pint	pt	33.600 cu in	0.550 liter

Length

mile	mi	5280 ft, 1760 yd	1.6909 kilometers
rod	rd	5.50 yd, 16.5 ft	5.029 meters
yard	yd	3 feet, 36 inches	0.9144 meter
foot	ft or '	12 in, 0.333 yd	30.480 centimeters
inch	in or "	0.083 ft, 0.027 yd	2.540 centimeters

Area

square mile	sq mi	640 acres	2.590 square kilometers
acre		43,560 sq ft	4047 square meters
square rod	sq rd	3.25 sq yd	25.293 square meters
square yard	sq yd	9 sq ft	0.836 square meter
square foot	sq ft	144 sq in	0.093 square meter
square inch	sq in	0.007 sq ft	6.451 square centimeters

Volume

cubic yard	cu yd	27 cu ft	0.765 cubic meter
cubic foot	cu ft	1728 cu in	0.028 cubic meter
cubic inch	cu in	0.00058 cu ft	16.387 square centimeters

Speed

inches per second	in/sec	0.0167 in/min	2.54 centimeters/second
inches per minute	in/min	0.083 ft/min	2.54 centimeters/minute
feet per second	ft/sec	60 ft/min	30.480 centimeters/second
feet per minute	ft/min	720 in/sec	30.480 centimeters/minute

Mixing

ounces per gallon dry to wet	oz/gal		7.49 grams/liter
ounces per gallon wet to wet	oz/gal		7.81 milliliters/liter

Appendix Five
Additional Reading & Reference

Albrecht, Karl. *At America's Service*. Homewood, Ill.: Dow Jones Irwin, 1988.

Albrecht, Karl; and Ron Zemke. *Service America*. Homewood, Ill.: Dow Jones Irwin, 1985.

Bhote, Keki R. *Strategic Supply Management*. New York: American Management Association, 1989.

World Class Quality. New York: Amacom, 1991.

Byham, William C.; with Jeff Cox. *Zapp, The Lightning of Empowerment*. New York: Harmony Books, 1990.

Clifford, Donald K., Jr.; and Richard E. Cavanagh. *The Winning Performance*. New York: Bantam Books, 1985.

Covey, Stephen R. *The Seven Habits of Highly Effective People*. New York: Simon & Schuster, 1989.

Crosby, Philip B. *Quality Is Free: The Art of Making Quality Free*. New York: McGraw-Hill, 1979.

Quality Without Tears. New York: New American Library, 1984.

Davidow, William; and Bro Uttal. *Total Customer Service—The Ultimate Weapon.* New York: Harper and Row, 1989.

Deming, W. Edwards. *Japanese Methods for Productivity and Quality.* Washington, D.C.: George Washington University, 1981.

Quality Productivity and Competitive Position. Cambridge, Mass.: Massachusetts Institute of Technology, 1982.

Out of the Crisis. Cambridge, Mass.: Massachusetts Institute of Technology, 1986.

Feigenbaum, Armand V. *Total Quality Control.* New York: McGraw Hill, 1983.

Fukunda, Ryuji. *Managerial Engineering-Techniques for Improving Quality and Productivity in the Workplace.* Stamford, Conn.: Productivity, Inc., 1983.

Gelsanliter, David. *Jump Start.* New York; Farrar, Straus & Giroux, 1990.

Goldratt, Eliyahu M.; and Jeff Cox. *The Goal.* New York: North River Press, 1984.

Greene, James H. *Production and Inventory Control.* New York: Business One Irwin, 1974.

Gunn, Thomas G. *Manufacturing for Competitive Advantage.* New York: Harper Business, 1987.

Hall, Robert W. *Attaining Manufacturing Excellence.* New York: Business One Irwin, 1987.

Harmon, Roy L.; and Leroy D. Peterson. *Reinventing the Factory,* New York: The Free Press, 1990.

Imai, M. *Kaizen.* New York: Random House, 1986.

Ishikawa, Kaoru. *Guide to Quality Control.* Lanham, Md.: UNIPUB, 1976.

What Is Total Quality Control, The Japanese Way. Englewood Cliffs, N.J.: ASQC Quality Press, Prentice Hall, 1985.

Juran, Joseph M. *Managerial Breakthrough: A New Concept of the Manager's Job.* New York: McGraw-Hill, 1964.

Quality Control Handbook. New York: McGraw-Hill, 1979.

Juran on Planning for Quality. New York: The Free Press, 1988.

Kami, Michael J. *Trigger Points.* New York: McGraw-Hill, 1988.

Mizuno, Shigero. *Company Wide Total Quality Control.* New York: Kraus International, 1987.

Monden, Yasuhiro. *Toyota Production System.* Atlanta: Institute of Industrial Engineers, 1983.

Applying Just-In-Time: The American/Japanese Experience. Norcross, Ga.: Industrial Engineering and Management Press, 1986.

Nemoto, Masao. *Total Quality Control for Mangement—Strategies and Techniques from Toyota and Toyota Gosei.* Englewood Cliffs, N.J.: Prentice Hall, 1987.

Ouchi, William G. *Theory Z.* Reading, Mass.: Addison-Wesley, 1981.

Peters, Thomas; and N. Austin. *A Passion for Excellence.* New York: Random House, 1985.

Peters, Thomas J.; and Robert H. Waterman, Jr. *In Search of Excellence: Lessons from America's Best Run Companies.* New York: Harper and Row, 1982.

Pinchot, G. *Intrapreneuring.* New York: Harper and Row, 1985.

Reid, Peter C. *Well Made in America.* New York: McGraw-Hill, 1990.

Schonberger, Richard J. *Japanese Manufacturing Techniques: Nine Hidden Lessons in Simplicity.* New York: The Free Press, 1982.

World Class Manufacturing-The Lessons of Simplicity Applied. New York: Macmillan, Inc., 1986.

Shores, Richard A. *Survival of the Fittest.* Milwaukee, Wis.: ASQC Quality Press, 1988.

Singo, Shigeo. *Zero Quality Control: Source Inspection and the Poka-Yoke System.* Cambridge, Mass.: Productivity Press, 1985.

Suzaki, Kiyoshi. *The New Manufacturing Challenge, Techniques for Continuous Improvement*. New York: The Free Press, 1987.

von Oech, Roger. *A Whack on the Side of the Head*. New York: Warner Books, 1990.

Waitley, Dennis. *The Psychology of Winning*. New York: Berkley Books, 1979.

Walton, Mary. *The Deming Management Method*. New York: Dodd, Mead and Co., 1986.

Womack, James P.; Daniel T. Jones; and Daniel Roos. *The Machine That Changed the World*. New York: Rawson Associates, 1990.

Malcolm Baldrige Award Criteria

Listed below are seven categories used by the Malcolm Baldrige National Quality Award committee for determining outstanding companies. This reference is provided for understanding the importance of total quality management in all aspects of the business.

1992 Examination Categories/Items   **Point Values**

1.0 Leadership 90
    1.1 Senior Executive Leadership 45
    1.2 Management for Quality 25
    1.3 Public Responsibility 20

2.0 Information and Analysis 80
    2.1 Scope and Management of Quality and
        Performance Data and Information 15
    2.2 Competitive Comparisons and Benchmarks 25
    2.3 Analysis and Uses of Company-Level Data 40

3.0 Strategic Quality Planning 60

Appendix Seven
Types of Inventory

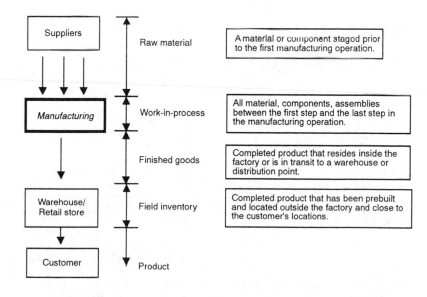

Suppliers	Raw material	A material or component staged prior to the first manufacturing operation.
Manufacturing	Work-in-process	All material, components, assemblies between the first step and the last step in the manufacturing operation.
	Finished goods	Completed product that resides inside the factory or is in transit to a warehouse or distribution point.
Warehouse/ Retail store	Field inventory	Completed product that has been prebuilt and located outside the factory and close to the customer's locations.
Customer	Product	

Index

X

FORECASTING SYSTEMS FOR OPERATIONS MANAGEMENT
Stephen A. Del urgio and Carl D. Bhame

Understand and implement practical, theoretically sound, and comprehensive forecasting systems. *Forecasting Systems for Operations Management* will assist you in moving products, materials, and timely information through your organization. It's the most comprehensive treatment of forecasting methods available for automated forecasting systems.
1-55623 04-0 $44.95

COMMON SENSE MANUFACTURING
Becoming a Top Value Competitor
James A. Gardner

Gardner details how you can integrate your manufacturing process and transform your company into a world-class competitor...even if you have limited resources. This common-sense approach to quality and service enhancement pays off in faster employee acceptance and systems applications. Using Gardner's straightforward planning suggestions, you can reduce floor space through better plan layout and gradually eliminate work-in-progress inventory.
1-55623-527-5 $34.95

PURCHASING STRATEGIES FOR TOTAL QUALITY
A Guide to Achieving Continuous Improvement
Greg Hutchins

Shows how companies can use purchasing to help meet and exceed customer demands. Hutchins details how elevating the purchasing person from order taker to a partner in the manufacturing process can be vital to the success of quality improvement efforts. He shows you how to establish continuous improvement strategies with suppliers and infuse quality throughout your organization.
1-55623-380-9 $42.50

COMPUTER INTEGRATED MANUFACTURING
Guidelines and Applications from Industrial Leaders
Steven A. Melnyk and Ram Narasimhan

Melnyk and Narasimhan offer a strategic, management-based approach for understanding and implementing CIM in your operation. A CIM book written from the manager's point of view rather than that of a technical expert, *Computer Integrated Manufacturing* is a clear, easy-to-understand guide. It shows you how to develop a strong link between strategy and CIM structure and build a competitive advantage over your business rivals.
1-55623-538-00 $45.00

MANUFACTURING PLANNING AND CONTROL SYSTEMS
Third Edition
Thomas E. Vollmann, William L. Berry, and D. Clay Whybark

In the Third Edition of *Manufacturing Planning and Control Systems*, state-of-the-art concepts and proven techniques are combined to offer a practical solution to enhancing the manufacturing process. Each of the book's central themes—Master Planning, Material Requirements Planning, Inventory Management, Capacity Management, Production Activity Control, and Just-in-Time—has been updated to reflect the newest ideas and practices.
1-55623-608-5 $55.00